Dissertation

Interpolating Scaling Vectors and Multiwavelets in $\mathbb{R}^d$

A Multiwavelet Cookery Book

Karsten Koch

2006

Bibliografische Information der Deutschen Nationalbibliothek

Die Deutsche Nationalbibliothek verzeichnet diese Publikation in der
Deutschen Nationalbibliografie; detaillierte bibliografische Daten sind
im Internet über http://dnb.d-nb.de abrufbar.

ISBN 978-3-8325-1489-1

Logos Verlag Berlin
Comeniushof, Gubener Str. 47,
10243 Berlin
Tel.: +49 030 42 85 10 90
Fax: +49 030 42 85 10 92
INTERNET: http://www.logos-verlag.de

Interpolating Scaling Vectors and Multiwavelets in $\mathbb{R}^d$

Dissertation
zur
Erlangung des Doktorgrades
der Naturwissenschaften
(Dr. rer. nat.)

dem

Fachbereich Mathematik und Informatik
der Philipps–Universität Marburg

vorgelegt von

Karsten Koch
aus Marburg/Lahn

Marburg/Lahn Oktober 2006

Vom Fachbereich Mathematik und Informatik
der Philipps–Universität Marburg als Dissertation
angenommen am: 08.12.2006

Erstgutachter: Prof. Dr. Stephan Dahlke, Philipps–Universität Marburg

Zweitgutachter: Prof. Dr. Peter Maaß, ZeTeM, Universität Bremen

Drittgutachter: Prof. Dr. Gabriele Steidl, Universität Mannheim

Tag der mündlichen Prüfung: 14.12.2006

Acknowledgements

First of all, I would like to express my gratitude to my referees, Prof. Dr. Stephan Dahlke, Prof. Dr. Peter Maaß, and Prof. Dr. Gabriele Steidl, for their willingness to wade through this thesis which surely contains some notational obstacles and much to often the prefix "multi". Moreover, Stephan did not only supervise this thesis but also spent many hours on discussing the various problems I stumbled into. Without him, this work would neither have attained its content nor its final shape. In addition, special thanks go to my colleagues at the AG Numerik/Wavelet Analysis for a very nice atmosphere and for enlightening though not necessarily mathematical discussions. In particular, I am indebted to Thorsten Raasch who did not only play a crucial role in the beginning of my time in Marburg but has always been inclined to share the blessings of his impressive mnemonic capability as well. I would like to thank OStD Winfried Damm who taught me that doing mathematics means more than just shuffling some symbols and therefore raised my interest in mathematics. Furthermore, I want to express my gratitude to Prof. Dr. Joachim Ohser who encouraged me to do this doctorate and, by the way, changed my view of applied mathematics to a great extent. I also feel grateful to the Deutsche Forschungsgemeinschaft which has supported my whole stay in Marburg by means of the Grants Da 360/4–(1–3).

Finally, I wish to thank my parents for encouraging and supporting all my endeavors, and last but not least, of course, my very special thanks go to Sandra for her great patience and the emotional support without which this project would never have been completed.

Zusammenfassung

Wavelets sind spezielle Basen des $L_2(\mathbb{R})$, die durch dyadische Dilatation und ganzzahlige Translation einer einzigen Funktion, des sogenannten Mutterwavelets, entstehen. Der große Vorteil von Waveletbasen ist, dass sie eine skalenweise Approximation von quadratintegrablen Funktionen zulassen, wobei jeder Skalenübergang als Detailgewinn interpretiert werden kann. Daher haben sich Wavelets innerhalb der letzten zwei Jahrzehnte zu einem wertvollen Hilfsmittel sowohl in der angewandten als auch in der reinen Mathematik entwickelt. So bilden Wavelets beispielsweise einen festen Bestandteil des JPEG2000-Standards zur Bilddatenkompression, werden aber gleichzeitig auch in der Approximationstheorie zur Charakterisierung verschiedener Funktionenräume genutzt.

Heutzutage werden Wavelets im Allgemeinen mittels einer sogenannten Multiskalenanalyse konstruiert. Diese wiederum wird von einer einzelnen quadratintegrablen Funktion erzeugt, der Skalierungsfunktion. Da nahezu sämtliche Eigenschaften eines Wavelets von diesem Generator abhängen, wurde und wird die Konstruktion dieser Skalierungsfunktionen in der Literatur eingehend behandelt. Dabei stellt sich heraus, dass das klassische Waveletkonzept einigen Beschränkungen unterliegt. Es lässt sich beispielsweise zeigen, dass keine kompakt getragene interpolierende Skalierungsfunktion mit orthogonalen Translaten existiert, welche gleichzeitig stetig ist. Diese Eigenschaften sind jedoch inbesondere für Anwendungszwecke sehr erstrebenswert. So erlaubt die Orthogonalität eines Generators die Konstruktion einer orthogonalen Waveletbasis, während die Interpolationseigenschaft zu einem Shannon-artigen Abtasttheorem führt, welches die Berechnung der Waveletentwicklung einer Funktion maßgeblich erleichtert.

Ein möglicher Ansatz zur Umgehung dieser Einschränkungen ist der Versuch, die obigen Forderungen etwas abzuschwächen. So verzichtet man häufig auf Orthogonalität zugunsten einer schwächeren Biothogonalitätsbedingung, d.h., anstelle einer orthogonalen Skalierungsfunktion betrachtet man zwei zueinander duale Skalierungsfunktionen, welche zu biorthogonalen Waveletbasen führen. Jedoch birgt auch dieses Konzept einige Nachteile. Es zeigt sich in vielen Konstruktionen, die diesen Ansatz verfolgen, dass für gewöhnlich starke Eigenschaften des primalen Generators mit schwachen Eigenschaften des dualen Generators einhergehen.

Aus diesem Grund befassen wir uns in der vorliegenden Arbeit mit Multiwavelets, einer Verallgemeinerung des klassischen Waveletkonzeptes, welche deutlich mehr Spielraum zur Konstruktion übrig lässt. Multiwaveletbasen werden, im Gegensatz zur ihren klassischen Verwandten, nicht nur von einem einzelnen sondern von mehreren Mutterwavelets erzeugt, die für gewöhnlich in einem Vektor angeordnet werden. Wie skalare Wavelets werden auch Multiwavelets zumeist mittels einer Multiskalenanalyse konstruiert, welche ihrerseits von einer vektorwertigen Funktion, dem Skalierungsvektor, generiert wird. Unser Hauptziel in dieser Arbeit ist die Konstruktion eben solcher Skalierungsvektoren und der dazugehörigen Multiwavelets in sowohl einer, als auch in mehreren Veränderlichen. Ein besonderes Augenmerk liegt dabei auf der Konstruktion interpolierender Skalierungsvektoren.

Für unseren multivariaten Ansatz verwenden wir das Konzept allgemeiner Skalierungsmatrizen, das eine natürliche Erweiterung des klassischen dyadischen Dilatationprinzips darstellt. Allerdings zeigt sich, dass der multivariate skalare Fall nahezu die gleichen Einschränkungen wie sein univariates Gegenstück aufweist. Aus diesem Grunde bietet sich hier ebenfalls der Übergang zu Multiwavelets an. Jedoch stellt die Konstruktion multivariater Multiwavelets eine gewisse Herausforderung dar, da bisher, im Gegensatz zum univariaten Fall, nicht bekannt ist, ob für jeden Skalierungsvektor ein zugehöriges Multiwavelet gefunden werden kann. Alles in allem lassen sich die Ziele dieser Arbeit in den folgenden grundlegenden Fragestellungen zusammenfassen:

(T1) Welches Potenzial steckt in dem vektorwertigen Ansatz? Lassen sich die Beschränkungen des skalaren Falles damit beheben?

(T2) Existieren Multiwavelets für jeden interpolierenden Skalierungsvektor? Gibt es vielleicht eine Art kanonisches Multiwavelet?

(T3) Sind diese Konzepte auch in der Anwendung von Nutzen?

Nach einer kurzen Diskussion des klassischen Waveletkonzeptes in Kapitel 2 und einer Einführung der grundlegenden Begriffe in Kapitel 3, welches insbesondere die Definition eines neuartigen Interpolationsbegriffes für Skalierungsvektoren in mehreren Veränderlichen beinhaltet, wenden wir uns der Beantwortung der obigen Fragen zu. Dafür entwickeln wir in Kapitel 4 zuerst einen systematischen Ansatz zur Konstruktion orthogonaler interpolierender Skalierungsvektoren in einer Veränderlichen, welche einen kompaktem Träger besitzen. Der resultierende Algorithmus erlaubt nicht nur die zur Zeit führenden Ergebnisse aus [109] zu reproduzieren, sondern gleichzeitig noch weitere Skalierungsvektoren zu konstruieren, die bei ansonsten gleichen Eigenschaften eine höhere Regularität aufweisen. Des Weiteren dient uns dieser univariate Zugang als eine Art Schablone für die Konstruktionsmethoden in den folgenden Kapiteln. So erweitern wir diesen

Ansatz in Kapitel 5 zu einen Algorithmus für die Konstruktion kompakt getragener orthogonaler interpolierender Skalierungsvektoren in mehreren Veränderlichen für Skalierungsmatrizen mit Determinante ± 2. Neben der expliziten Konstruktion einiger bivariater Beispiele entwickeln wir dort eine Regel, mit Hilfe derer sich problemlos geeignete Multiwavelets angeben lassen. In Kapitel 6 untersuchen wir den biorthogonalen Fall unter Hinzunahme verschiedener Symmetriebedingungen. Die Hauptergebnisse dieses Abschnittes lassen sich wie folgt zusammenfassen. Zuerst leiten wir einen Algorithmus zur Konstruktion biorthogonaler Paare kompakt getragener symmetrischer Skalierungsvektoren in mehreren Veränderlichen her, wobei die primalen Funktionen interpolieren. Außerdem zeigen wir, dass zu jedem dieser Paare in einer kanonischen Weise Multiwavelets konstruiert werden können. Abschließend geben wir einige bivariate Beispiele an. Im letzten Kapitel dieser Arbeit untersuchen wir die Anwendbarkeit der konstruierten Multiwavelets. Dabei beschränken wir uns auf das Gebiet der Bilddatenkompression, welches eine Standardanwendung für Wavelets darstellt.

Die Ergebnisse dieser Arbeit lassen sich in den folgenden Antworten auf die Fragen (T1)–(T3) zusammenfassen:

(T1) Sowohl der univariate Ansatz in Kapitel 4 als auch dessen multivariates Analogon in Kapitel 5 führen zu interpolierenden Skalierungsvektoren, welche neben einem kompakten Träger auch orthogonale Translate besitzen. Ferner sind die meisten der dort konstruierten Beispiele mindestens stetig oder sogar stetig differenzierbar, was im skalaren Fall nicht erreicht werden kann. Des Weiteren bieten die in Kapitel 6 konstruierten biorthogonalen Skalierungsvektoren ebenfalls einen Vorteil gegenüber ihren skalaren Verwandten. So besitzen insbesondere die dualen Skalierungsvektoren eine im Vergleich zum skalaren Fall deutlich gesteigerte Glattheit bei ansonsten identischen Eigenschaften. Diese Ergebnisse zeigen, dass das Multiwaveletkonzept wesentliche Vorteile gegenüber dem skalaren Fall bietet.

(T2) Zusätzlich zu jedem der oben genannten Algorithmen zur Konstruktion interpolierender Skalierungsvektoren entwickeln wir Methoden zur Konstruktion der dazugehörigen Multiwavelets. Im orthogonalen Fall, d.h. in den Kapiteln 4 und 5, besteht diese Methode aus einer einfachen Regel, die es erlaubt, geeignete Multiwavelets sofort anzugeben. Für den allgemeineren biorthogonalen Fall leiten wir in Kapitel 6 ein Verfahren her, welches für jedes Paar biorthogonaler Skalierungsvektoren mit kompaktem Träger ein Paar zugehöriger kanonischer Multiwavelets liefert, sofern der primale Skalierungsvektor unsere Interpolationsbedingung erfüllt.

(T3) Zuerst zeigen wir in Kapitel 7, dass interpolierende Skalierungsvektoren und die dazugehörigen Multiwavelets prinzipiell für die Anwendung geeignet sind, da sie eine weitere wesentliche Approximationseigenschaft seitens ihrer Filter besitzen. Außerdem weisen die in Kapitel 7 erzielten Kompressionsergebnisse darauf hin, dass unsere Multiwavelets auch für diese spezielle Anwendung nützlich sind. Allerdings hängen die Ergebnisse der multivariaten Multiwavelets vom Zusammenspiel der Bild- und Waveletcharakteristiken ab. Unsere univariten Multiwavelets hingegen liefern uneingeschränkt gute Resultate.

"First, Catch Your Hare ..."

Hannah Glasse, *The Art of Cookery Made Plain and Easy (1747)*

Contents

List of Figures

List of Tables

Chapter 1

Introduction

Within the last two decades, wavelet analysis has become a very powerful tool in applied mathematics. Wavelet algorithms have been successfully applied in signal analysis and compression as well as in numerical analysis, geophysics, meteorology and in many other fields. Moreover, due to their strong analytical features, wavelets can be utilized in pure mathematics as well. For example, their ability to characterize certain function spaces has made wavelets an esteemed tool in approximation theory.

Similar to the wide range of applications, also the roots of wavelet theory can be found in various mathematical fields. Some of the most renown predecessors of wavelets are the time-frequency atoms introduced by Gabor in [48] along with the windowed Fourier transform. There, modulated and translated wave functions, typically Gaussians, are used to measure time-localized frequency contents of signals or functions. However, due to the fixed support of the Gabor atoms, the provided time resolution is fixed as well. For this reason, in the early 1980's the geophysicist Morlet proposed to utilize scaling instead of modulation, i.e., to use atoms of the form

$$a^{-1}\psi\left(\frac{x-b}{a}\right), \qquad a > 0,\, b \in \mathbb{R},$$

where ψ is a suitable wave function or wavelet. By choosing small scaling parameters a, one obtains an arbitrarily high resolution in time or space, respectively, and therefore wavelets can be used as a kind of mathematical microscope. Together with the results of Grossmann, Morlet's approach climaxed in the invention of the continuous wavelet transform, cf. [54]. Although this continuous transform has proven to be proficient in various applications, from the numerical point of view it is often more desirable to deal with a discrete transform. Hence, as a matter of course, the question arised how to discretize this continuous wavelet setting. The first results in this direction were obtained in [38]. There, it was shown that for a proper choice of the wavelet ψ and some parameters $a_0 > 1$ and $b_0 > 0$, a sampling of the form

$$a_0^{j/2}\psi(a_0^j x - b_0 k), \qquad j, k \in \mathbb{Z},$$

generates a frame, i.e., a system which provides a stable decomposition of functions in $L_2(\mathbb{R})$. Moreover, later on it turned out that the atoms obtained by such a sampling can even constitute an orthonormal basis of $L_2(\mathbb{R})$. These results along with aspects of several other fields like, e.g., signal processing and operator theory, crystallized in the notion of multiresolution analyses introduced by Mallat and Meyer in [87, 91]. Although there exist some earlier constructions of wavelet bases, cf. [55, 89, 115], the concept of multiresolution analyses has been the first systematical approach to the construction of wavelets. Since the advent of this concept, the literature on wavelets has dramatically increased such that nowadays there is an enormous number of papers on this topic. Hence, it is impossible to give a complete list of references. Instead, we refer to the well-known textbooks [23, 37, 86, 88, 91, 118] and the just published compilation [66] which contains a fine selection of the fundamental papers in wavelet theory.

Mostly, the interest has centered around the dyadic wavelet case, i.e., a function or signal is analyzed by the dyadic dilates and integer translates of one mother wavelet ψ such that the $2^{j/2}\psi(2^j \cdot - k)$, $j, k \in \mathbb{Z}$, constitute an orthonormal basis of $L_2(\mathbb{R})$. Hence, one obtains a multiscale representation of functions in $L_2(\mathbb{R})$ with respect to the scale j. The underlying multiresolution analysis is generated by one single function φ called the scaling function which satisfies a so-called refinement equation

$$\varphi(x) = \sum_{\beta \in \mathbb{Z}} a_\beta \varphi(2x - \beta) \qquad \text{for almost all } x \in \mathbb{R} \tag{1.1}$$

with a mask $(a_\beta)_{\beta \in \mathbb{Z}}$. This generator plays a fundamental role in the construction of wavelets, since almost all properties of wavelets are inherited from the scaling function φ. For application purposes, one is usually interested in good decay properties of φ. Hence, compactly supported scaling functions are appreciated most. Furthermore, it is often convenient to use interpolating scaling functions, i.e., φ is at least continuous and satisfies

$$\varphi(\beta) = \delta_{0,\beta} \qquad \text{for all } \beta \in \mathbb{Z}. \tag{1.2}$$

The main benefit of interpolating scaling functions is that they provide a Shannon–like sampling theorem which facilitates the computation of the wavelet decomposition of a function, cf. Section 3.1.2 for details. However, then it turns out that the classical wavelet setting is somewhat restricted. It has been shown in Chapter 6 of [37], see also [120], that the Haar function, i.e., the characteristic function of the unit interval, is the only compactly supported scalar generator of a multiresolution analysis which is orthonormal and satisfies (1.2). Moreover, as proven in Chapter

8 of [37], there exists only one compactly supported scaling function which is symmetric — again, it is the Haar function. Hence, the scalar setting fails to provide certain desirable properties simultaneously.

To bypass this lack of flexibility, several approaches have been proposed. One possibility is to switch over to the already above mentioned weaker concept of wavelet frames, i.e., the dilates and translates of one or more mother wavelets do no longer constitute an orthonormal basis but a redundant system which provides a stable decomposition of functions in $L_2(\mathbb{R})$, see [17] and the references therein for details. Nevertheless, in many applications the strong basis properties are very desirable. Then, another possible way to increase flexibility is to use biorthogonal wavelet bases. Instead of one orthonormal wavelet basis, one employs a dual pair of stable wavelet bases which are biorthogonal. However, as we will see later on, also this setting bears some limitations. Another very promising approach to overcome the above restrictions is to use multiwavelets. These appear as a natural generalization of the scalar wavelet setting, and thus they can either constitute orthonormal or biorthogonal bases of $L_2(\mathbb{R}^d)$. Instead of one mother wavelet, one considers several mother wavelets which are organized in vectors. Hence, one often refers to this setting as the vector setting.

The notion of multiwavelets goes back to the early to mid-1990's. The synonymously used terms r–vector multiresolution analysis and multiresolution analysis of multiplicity $r > 1$ have been introduced rather independently in [5, 39, 51, 67], and the first orthonormal multiwavelet bases have been constructed there. As in the scalar case, vector multiresolution analyses are usually obtained by means of a generator, a vector valued function called scaling vector which satisfies a refinement equation similar to (1.1) with a matrix valued mask, cf. Section 3.1.1. The potential of the vector approach has become evident with the construction of continuous symmetric scaling vectors with compact support in [49]. Thenceforth, multiwavelets have evolved into a field of current mathematical research, cf. [26, 94, 99, 111, 114] and the references therein. Furthermore, in [109] the notion of interpolating or cardinal scaling vectors has been introduced by means of an interpolation property similar to Equation (1.2). In addition, interpolating scaling vectors with compact support have been constructed there which are continuous or, moreover, continuously diferentiable. Yet again, the limitations of the scalar setting are overcome by the vector approach.

All the generalizations discussed so far aim at increasing the flexibility of the classical wavelet setting. Hence, they all focus on systems which allow a stable decomposition of functions in $L_2(\mathbb{R})$. On the other hand, in many applications multivariate functions occur. Therefore, almost since the advent of wavelet theory, there have been attempts to generalize the univariate wavelet concept to multiple dimensions. A straightforward approach is a one-to-one translation of the univariate setting to functions in $L_2(\mathbb{R}^d)$, i.e., one searches for bases of $L_2(\mathbb{R})$ which are generated by dyadic dilates and integer translates of some mother wavelets. However, then it turns out that a multiresolution analysis approach necessarily leads to $2^d - 1$ mother wavelets. As a consequence, the multiscale decomposition of functions in $L_2(\mathbb{R}^d)$ becomes somewhat coarse, since for each scale j there is a huge number of distinct basis functions. One possibility to bypass this problem is to use a more general notion of scaling, i.e., to utilize expanding integer scaling matrices. It has been shown in [24, 90] that the number of mother wavelets corresponding to a multiresolution analysis with scaling matrix M is $|\det(M) - 1|$. Therefore, choosing a scaling matrix with determinant ± 2 allows the construction of wavelet bases generated by one single mother wavelet only. Another advantage of scaling matrices is that, in contrast to a uniform dilation parameter, they can possess certain rotation or reflection properties, and thus the corresponding wavelets can show distinct directional features on each scale.

Nevertheless, also for the multivariate setting, scalar wavelets show some restrictions. For example, one can show that for the case $|\det(M)| = 2$ there exists no compactly supported orthonormal generator of a multiresolution analysis which is continuous and satisfies an interpolation condition similar to (1.2) simultaneously, cf. Theorem 5.1.4 in the present work. Therefore, in the past research focused mainly on the biorthogonal case, see, e.g., [30, 31, 42, 43, 62, 63, 76, 101] and the references therein. However, all these approaches have in common that strong properties of the primal wavelets are accompanied by weak properties of the dual wavelets, i.e., a smooth primal wavelet with decent support corresponding to an interpolating scaling function usually leads to dual wavelets with either poor regularity properties or a huge support size. Thus, once again, multiwavelets seem to provide an adequate setting to overcome these restrictions. As a consequence, in recent years, the field of multivariate multiwavelets and scaling vectors has been extensively studied, see, e.g., [47, 52, 75, 76, 78, 96, 103] and the survey articles [12, 18]. Of course, this list can not be exhaustive.

Current constructions of multivariate scaling vectors are frequently settled in the context of stationary vector subdivision. Subdivision schemes evolved from the field of computer-aided geometrical design almost at the same time as wavelets appeared, and both concepts profited vastly from mutual influences, see [14, 34] for a survey on this topic. In particular, interpolating generators of multiresolution analyses have been important to both fields since they lead to interpolatory subdivision schemes. Hence, it is not surprising that the initial notion of interpolating scaling vectors in higher dimensions stems from the subdivision context. In [20] and in [28] first concepts of interpolatory vector subdivision schemes have been introduced which are closely related to interpolating scaling vectors. Furthermore, multivariate scaling vectors wich satisfy a Hermite interpolation condition have been obtained in [64]. However, since all these approaches aim at designing vector subdivision schemes, the problem of constructing some corresponding multiwavelet bases has not been addressed.

In the present work, we tackle the problem of constructing interpolating scaling vectors in $\mathbb{R}^d$ from a more wavelet-related point of view. Motivated by the results of Selesnick in [109], we propose an interpolation property for multivariate scaling vectors which can be considered as a natural extension of its scalar counterpart (1.2). Furthermore, it turns out that our interpolation condition does perfectly fit into both notions of interpolatory vector subdivision schemes introduced in [20, 28]. Nevertheless, the main aim of this work is the construction of multivariate interpolating scaling vectors in the wavelet context. Therefore, we intend to address the following fundamental topics:

(T1) Does the vector approach provide enough flexibility to overcome the restrictions of the scalar setting? To which extent can the vector setting be exploited?

(T2) Can we always find multiwavelets corresponding to interpolating scaling vectors? Do there exist any canonical multiwavelets?

(T3) Are these concepts suitable for application purposes?

Since we want to fathom the potential of the vector setting, our construction principle has to be very systematical. Hence, we do not only want to obtain some good results but find optimal ones in sets of scaling vectors with similar properties. Therefore, we choose a bottom-up approach which, when fed with the proper input data, can lead to any possible scaling vector. Thus, our construction methods do always contain an optimization part. However, most of the above mentioned approaches fail to meet our demands, and therefore can not be used as templates for our construction as we shall now explain. On one hand, many methods simply can not be generalized to the multivariate vector case. For example, scalar methods are often based on a factorization of the symbol, i.e., a Laurent series which is determined by the mask of the scaling function. The more complex algebraic structure of the vector setting prohibits such an approach. On the other hand, most construction methods for both the scalar and the vector case do not match our criteria concerning systematicality. Most of them employ some deep insights or highly sophisticated tricks which allow to obtain some scaling functions or vectors with very nice properties. But these results are often somewhat singular or isolated, i.e., they are in general not embedded into a larger context which makes it hard to determine whether they are optimal in some sense or not. A good example for such an approach is the one given by Selesnick in [109]. There, very nice interpolating scaling vectors with decent additional properties are obtained by means of a Gröbner bases approach. Nevertheless, we will see in the sequel that some of the results presented in [109] are not optimal in the sense that there exist scaling vectors with exactly the same properties but a higher regularity.

This thesis is organized as follows. In Chapter 2, we briefly recall the classical univariate scalar wavelet setting to introduce the basic concepts without having to deal with too much irritating notational complexity. The next chapter is devoted to the general setting. There, we introduce the notion of multivariate scaling vectors and multiwavelets and state our new interpolation condition. In addition, we discuss the relation of our setting to the field of interpolatory subdivision schemes. Then, in Chapter 4, we develop a systematical approach for the construction of univariate scaling vectors which are interpolating, orthonormal, and compactly supported simultaneously. This method is intended as a template for the multivariate construction methods in the following chapters. Although our approach is quite different to the one of Selesnick in [109], we partially reproduce the examples constructed there but, in addition, do also obtain some more regular examples. Moreover, we put some results stated in [109] on a sound mathematical foundation. The main part of this chapter has already been published in [79]. In Chapter 5, our univariate approach is generalized to the multivariate setting. Hence, we state an algorithm for the construction of orthonormal interpolating scaling vectors with compact support for scaling matrices with determinant ± 2. This approach is substantiated by some bivariate examples. Furthermore, we present a surprisingly simple method to obtain some corresponding multiwavelets with nice additional properties. These results have been published in [80]. Chapter 6 is devoted to the construction of multivariate interpolating scaling vectors with compact support which possess certain symmetry properties. There, we leave the orthonormal setting and focus on the biorthogonal case. Our main result in this chapter is a systematic method which does not only lead to biorthogonal pairs of symmetric interpolating scaling vectors but covers the scalar case as well. Moreover, we address the problem of finding multiwavelets and end up with an algorithm which leads to canonical multiwavelets corresponding to arbitrary interpolating scaling vectors. These results have already been summarized in [81]. Finally, in Chapter 7, we study the suitability of our results for application purposes. As a test application, we choose image compression, where we compare our scaling vectors with several well-established wavelets. In the end, our findings lead to the following positive answers to the questions (T1)–(T3):

(T1) The construction methods derived in Chapters 4 and 5 lead to orthonormal interpolating scaling vectors with compact support which, in most cases, are continuous or even continuously differentiable. Moreover, the biorthogonal scaling vectors obtained in Chapter 6 do always possess better properties than their scalar relatives in terms of regularity per support size. Thus, all the restrictions of the scalar setting can be overcome by the vector concept. Furthermore, our algorithms lead to optimal scaling vectors in the sense that they possess a maximum smoothness.

(T2) In Sections 4.3 and 5.3, respectively, we derive a simple but effective trick which enables us to obtain some multiwavelets corresponding to orthogonal interpolating scaling vectors in a very effortless way. Additionally, in Section 6.3, we develop a method to compute some multiwavelets for arbitrary dual pairs of scaling vectors whenever their masks are finitely supported and the primal scaling vector is interpolating. Since this approach is of a very fundamental nature, the obtained multiwavelets can be considered as canonical ones.

(T3) First of all, we show in Section 7.1.3 that interpolating scaling vectors are balanced, i.e., they possess an approximation property which is very important for application purposes. In addition, from the image compression results obtained in Chapter 7, we observe that the multivariate wavelets and multiwavelets constructed within this work perform very well at least for a certain class of images. Moreover, the univariate multiwavelets show very good results in all cases. Hence, interpolating scaling vectors seem to be well-suited for application purposes.

Chapter 2

Appetizer: The Classical Setting

Although this work is mainly concerned with the construction of multiwavelets, we give a brief introduction to the classical univariate wavelet setting at first. This enables us to introduce the basic ideas and to motivate our approach while, for now, sparing the reader the task of wading through complicated technical and notational details. First of all, we briefly recall the basic definitions of wavelets and multiresolution analyses and get into some details concerning their construction. This immediately leads us to the notion of refinability and so-called scaling functions. Afterwards, we address the question which properties of wavelets are desirable for application purposes. Finally, we give a short introduction to the field of stationary subdivision schemes which is closely related to the concept of refinable functions.

The following survey has been composed from Chapter 5 in [37] and from the Chapters 1–3 in [118]. For a more detailed discussion we refer to these textbooks. A comprehensive treatment of stationary subdivision schemes can be found in [14].

2.1 Basic Concept

For simplicity, we we start with the univariate setting. A *wavelet* is a function $\psi \in L_2(\mathbb{R})$ such that the collection

$$\left\{ 2^{\frac{j}{2}} \psi(2^j \cdot - \beta) \,\middle|\, j, \beta \in \mathbb{Z} \right\} \tag{2.1}$$

is an orthonormal basis of $L_2(\mathbb{R})$. A very simple example is given by the *Haar wavelet*

$$\psi(x) = \begin{cases} 1, & 0 \leq x < \frac{1}{2}, \\ -1, & \frac{1}{2} \leq x < 1, \\ 0 & \text{otherwise.} \end{cases} \tag{2.2}$$

This very first orthonormal wavelet basis was invented nearly a century ago by Haar in [55]. Similar bare hands constructions, though more sopisticated, have been obtained, e.g., by Strömberg in [115] and by Meyer in [89], see also the textbooks [37, 118].

In [87, 91], Mallat and Meyer introduced the concept of multiresolution analyses which provides a natural framework for the construction of wavelets. A *multiresolution analysis* (MRA) is a sequence $(V_j)_{j \in \mathbb{Z}}$ of closed subspaces of $L_2(\mathbb{R})$ which satisfies:

(MRA1) $V_j \subset V_{j+1}$ for each $j \in \mathbb{Z}$,

(MRA2) $g(x) \in V_j$ if and only if $g(2x) \in V_{j+1}$ for each $j \in \mathbb{Z}$,

(MRA3) $\bigcap_{j \in \mathbb{Z}} V_j = \{0\}$,

(MRA4) $\bigcup_{j \in \mathbb{Z}} V_j$ is dense in $L_2(\mathbb{R})$, and

(MRA5) there exists $\varphi \in L_2(\mathbb{R})$ such that $\{\varphi(x-\beta) \mid \beta \in \mathbb{Z}\}$ is an orthonormal basis in V_0.

The function φ in (MRA5) is called a *scaling function* or *generator* of the MRA. Condition (MRA2) implies that for each $j \in \mathbb{Z}$ the set $\{2^{j/2}\varphi(2^j x - \beta) \mid \beta \in \mathbb{Z}\}$ is an orthonormal basis for V_j. Thus, due to the nestedness (MRA1) of the V_j, φ has to satisfy the *refinement equation*

$$\varphi(x) = \sum_{\beta \in \mathbb{Z}} a_\beta \varphi(2x - \beta) \qquad \text{for almost all } x \in \mathbb{R}, \tag{2.3}$$

where the *mask* $(a_\beta)_{\beta \in \mathbb{Z}}$ is determined by the relation $a_\beta = 2\langle \varphi, \varphi(2 \cdot -\beta)\rangle$. Here, and in the following, $\langle \cdot, \cdot \rangle$ denotes the usual L_2 inner product.

Given a multiresolution analysis, to obtain a wavelet one proceeds as follows. First of all, for every $j \in \mathbb{Z}$ define the space W_j to be the orthogonal complement of V_j in V_{j+1}. Thus, we have

$$V_{j+1} = V_j \oplus W_j$$

and

$$W_j \perp W_{j'} \quad \text{if} \quad j \neq j'.$$

Therefore, (MRA3) and (MRA4) imply

$$L_2(\mathbb{R}) = \bigoplus_{j \in \mathbb{Z}} W_j,$$

such that $L_2(\mathbb{R})$ decomposes into mutually orthogonal subspaces. Furthermore, the spaces W_j inherit the property (MRA2) from the V_j, i.e., for each $j \in \mathbb{Z}$ we have

$$g \in W_j \quad \text{if and only if} \quad g(2\cdot) \in W_{j+1}. \tag{2.4}$$

Hence, if we find a function $\psi \in W_0$ such that $\{\psi(\cdot - \beta) \mid \beta \in \mathbb{Z}\}$ is an orthonormal basis of W_0, then for each $j \in \mathbb{Z}$ the set $\{2^{j/2}\psi(2^j \cdot -\beta) \mid \beta \in \mathbb{Z}\}$ constitutes an orthonormal basis for W_j. Since all the W_j are orthogonal, Equation (2.4) implies that ψ leads to an orthonormal wavelet basis of $L_2(\mathbb{R})$. Thus, given an MRA, the task of finding a wavelet basis reduces to finding an orthonormal basis of W_0 which consist of the integer translates of one single function $\psi \in W_0$.

Now assume that we have a wavelet corresponding to an MRA. Since $W_0 \subset V_1$, there has to exist a sequence $(b_\beta)_{\beta \in \mathbb{Z}}$ such that

$$\psi(x) = \sum_{\beta \in \mathbb{Z}} b_\beta \varphi(2x - \beta) \qquad \text{for almost all } x \in \mathbb{R}. \tag{2.5}$$

As a consequence, our task reduces to finding the sequence $(b_\beta)_{\beta \in \mathbb{Z}}$. The following theorem shows that there exists a canonical choice for this sequence. Furthermore, it ensures the existence of a wavelet corresponding to an MRA. A proof can be found in Chapter 5.1 in [37].

Theorem 2.1.1. *Let $(V_j)_{j \in \mathbb{Z}}$ be an MRA, and let φ be the corresponding scaling function with mask $(a_\beta)_{\beta \in \mathbb{Z}}$. Then with*

$$b_\beta := (-1)^\beta a_{1-\beta}, \qquad \beta \in \mathbb{Z}, \tag{2.6}$$

Equation (2.5) defines a wavelet ψ associated to $(V_j)_{j \in \mathbb{Z}}$.

Remark 2.1.2. *Of course, the choice (2.6) is not the only possible choice which leads to an orthonormal wavelet basis of $L_2(\mathbb{R})$. However, it is commonly known that if the generator φ has compact support and ψ is desired to have compact support as well, then the only possible choice is*

$$b_\beta := (-1)^\beta a_{1-\beta+2N}$$

for some $N \in \mathbb{Z}$. Thus, Equation (2.5) implies that the wavelet ψ is uniquely determined up to translation by an integer which does not alter the corresponding wavelet basis.

Finally, we have to address the problem of how to obtain an MRA. Although there are some approaches which directly construct the spaces V_j, a more common practice is to start with the scaling function φ. Then it turns out that under certain more or less mild conditions, a function φ that satisfies a refinement equation of the form (2.3) generates an MRA. This task, i.e., the construction of suitable scaling functions in a somewhat more general setting, is the main aim of this work. Hence, we refer to the following chapters for a detailed discussion of this topic.

2.2   Desirable Properties

The main benefit of wavelets is that they provide a multiscale representation of functions in $L_2(\mathbb{R})$ as follows. Once we are given an orthonormal wavelet basis associated to the wavelet ψ, every function $f \in L_2(\mathbb{R})$ has an expansion

$$f = \sum_{j\in\mathbb{Z}} \sum_{\beta\in\mathbb{Z}} d_{j,\beta} 2^{\frac{j}{2}} \psi(2^j \cdot -\beta) \tag{2.7}$$

with

$$d_{j,\beta} := 2^{j/2} \langle f, \psi(2^j \cdot -\beta) \rangle. \tag{2.8}$$

Now, restricting this representation to $j < J$ for some $J \in \mathbb{Z}$, we obtain an approximation of f on the *scale* J, i.e., in the space V_J. Hence, we are enabled to study approximations of f on different scales which might reveal distinct features of f.

For application purposes the wavelet expansion of a function is desired to possess several properties. First of all, in order to be able to compute or, at least, to approximate the coefficients $d_{j,\beta}$, the wavelet ψ should decay reasonably fast. Thus, a compactly supported wavelet ψ is most desirable. On the other hand, for almost all application purposes one is compelled to find a finite approximation of f. Therefore, for a certain class of *nice* functions, the above representation (2.7) is desired to be *sparse*, i.e., many of the coefficients $d_{j,\beta}$ vanish or are at least very small in modulus. If the functions f as well as the wavelet ψ are compactly supported, then for each fixed scale j we obtain a finite number of coefficients $d_{j,\beta} \neq 0$ only. Thus, the behaviour of the $d_{j,\beta}$ for $j \to \pm\infty$ remains to be controlled. For $j \to -\infty$, this problem can be overcome by switching to the representation

$$f = \sum_{\beta\in\mathbb{Z}} c_{J',\beta} 2^{\frac{J}{2}} \varphi(2^J \cdot -\beta) + \sum_{j\geq J'} \sum_{\beta\in\mathbb{Z}} d_{j,\beta} 2^{\frac{j}{2}} \psi(2^j \cdot -\beta) \tag{2.9}$$

with $c_{J',\beta} := 2^{J'/2} \langle f, \varphi(2^{J'} \cdot -\beta) \rangle$ for some coarsest scale $J' \in \mathbb{Z}$. If in addition the scaling function φ has compact support, then there are only finitely many $c_{J',\beta} \neq 0$. To obtain control over the $d_{j,\beta}$ for $j \to \infty$, we first have to specify our idea of the niceness of a function. Usually, nice functions are assumed to possess a certain degree of smoothness, e.g., we may think of polynomials or functions which can be approximated well using polynomials. Then the key property of a wavelet which leads to a sparse representation of the form (2.9) is a high order of *vanishing moments*, i.e., there exists an integer $k \geq 1$ such that

$$\int_{\mathbb{R}} x^n \psi(x)\,\mathrm{d}x = 0 \qquad \text{for} \quad n = 0, \dots, k-1.$$

Assume f is k–times continuously differentiable. For a large j and $\beta \in \mathbb{Z}^d$, f has the Taylor expansion

$$
\begin{aligned}
f(x) = \; & f(2^{-j}\beta) + f'(2^{-j}\beta)(x - 2^{-j}\beta) \\
& + \ldots + f^{(k-1)}(2^{-j}\beta)\tfrac{(x-2^{-j}\beta)^{k-1}}{(k-1)!} + (x - 2^{-j}\beta)^k R(x)
\end{aligned}
\tag{2.10}
$$

for x near $2^{-j}\beta$ such that the remainder R is bounded on each compact set. Hence, if ψ is a compactly supported wavelet with k vanishing moments, we obtain

$$
d_{j,\beta} = \int_{\mathbb{R}} (x - 2^{-j}\beta)^k R(x) 2^{\frac{j}{2}} \psi(2^j x - \beta) \, dx
$$

since the first k terms in (2.10) vanish. R is bounded on the support of $\psi(2^j \cdot -\beta)$, therefore we have

$$
|d_{j,\beta}| \le C 2^{-j(k-1/2)} \int_{\mathbb{R}} |x|^k |\psi(x)| \, dx
$$

for some constant C depending on R. Hence, $|d_{j,\beta}|$ will become small for large j unless R is very large near $2^{-j}\beta$. Consequently, the representation (2.9) is likely to be sparse.

For arbitrary $f \in L_2(\mathbb{R}^d)$ the moral of the story is as follows. Since the wavelet coefficients (2.8) reflect the local characteristics of f, they are small wherever f is locally smooth, while singularities of f lead to large wavelet coefficients. Thus, from the wavelet point of view, nice functions may possess a certain number of isolated singularities but are smooth ortherwise to yield sparse representations. Fortunately, many real world signals and images do perfectly fit into this description which is one reason for the present popularity of wavelet algorithms in signal processing. On the other hand, these observations do also indicate that wavelet algorithms are unsuitable for nonsmooth functions which possess lots of singularities.

In addition to a certain number of vanishing moments, the wavelet itself is also desired to possess a certain degree of smoothness. Obviously, to approximate smooth functions the approximants should be reasonably regular as well. On the other hand, the regularity of a wavelet coupled with its decay properties has a strong impact on the vanishing moment order of a wavelet. A proof of the following theorem can be found in Chapter 3 of [118].

Theorem 2.2.1. *Assume ψ is a k–times continuously differentiable wavelet which satisfies*

$$
|\psi(x)| \le \frac{C}{(1 + |x|)^\alpha}
$$

for some constant C and $\alpha > k + 1$. Furthermore, assume that all derivatives of ψ are bounded on $\mathbb{R}$. Then ψ has k vanishing moments.

Thus, regularity of the wavelet yields some vanishing moments. However, for a high order of vanishing moments, a wavelet does not necessarily have to be very smooth. We will see in Chapter 3.2.2 that the order of vanishing moments of a wavelet is vastly influenced by the approximation properties of the corresponding MRA, and thus by the properties of the underlying scaling function φ.

Besides the vanishing moment order, regularity induces the ability of wavelets to characterize certain function spaces as well. However, since this is no major topic of this work, we briefly sketch this ability for the case of Sobolev spaces only. For an arbitrary $s > 0$, the *Sobolev space* $H^s(\mathbb{R}^d)$ is defined by

$$H^s(\mathbb{R}) := \left\{ f \in L_2(\mathbb{R}) \,\Big|\, \int_{\mathbb{R}} |\widehat{f}(\xi)|^2 (1 + |\xi|^s)^2 \, d\xi < \infty \right\}$$

where $\widehat{f}$ denotes the standard Fourier transform of f. Now, if an MRA is generated by a k–times continuously differentiable scaling function with compact support, then for $s < k$ the corresponding wavelet ψ allows the characterization

$$f \in H^s(\mathbb{R}^d) \quad \text{if and only if} \quad \sum_{j \in \mathbb{Z}} (1 + 2^{2js}) \sum_{\beta \in \mathbb{Z}} |d_{j,\beta}|^2 < \infty$$

with the wavelet coefficients $d_{j,\beta}$ of f as in Equation (2.8). Hence, the Sobolev regularity of a function can be determined by the decay properties of its wavelet coefficients. The above characterization property can also be obtained for weaker requirements on the generator which involve the approximation properties of the MRA. For a detailed discussion of this topic, we refer to [91].

In practice, for a given function $f \in L_2(\mathbb{R})$, the vanishing moments of a wavelet may not suffice to obtain a finite representation of the form (2.9). This problem can be bypassed by truncating the representation, i.e., by introducing a finest scale J. Hence, one assumes that $f \in V_J$ and thus obtains the representation

$$f(x) = \sum_{\beta \in \mathbb{Z}} c_{J,\beta} \varphi(2^J x - \beta), \qquad x \in \mathbb{R}. \tag{2.11}$$

Nevertheless, in wavelet-related applications one is usually interested in a multi-scale representation of f. This can be obtained by utilizing the *discrete wavelet transform* DWT, i.e., an algorithm with decent properties which allows the effort-less computation of the wavelet coefficients $d_{j,\beta}$ for $J' \leq j < J$. In addition, the DWT provides the coefficients $c_{J',\beta}$ for the scaling function on the coarsest level J'. We will give a detailed discussion of this topic in Section 7.1.2.

In any case, an efficient algorithm to compute the fine scale coefficients in (2.11) is needed. To this end, it is very desirable that the scaling function φ is *interpolating*, i.e., φ is continuous and satisfies

$$\varphi(\beta) = \delta_{0,\beta}, \quad \text{for all } \beta \in \mathbb{Z}. \tag{2.12}$$

Then, from (2.11) one immediately obtains $c_{J,\beta} = f(2^{-J}\beta)$ for $f \in V_J$. Interpolating scaling functions have various nice properties, e.g., if φ is compactly supported, then (2.12) implies that for arbitrary sequences $u := (u_\beta)_{\beta \in \mathbb{Z}}$ the mapping

$$u \rightarrow \sum_{\beta \in \mathbb{Z}} u_\beta \varphi(\cdot - \beta)$$

is injective, and thus φ has *linearly independent integer translates*. Furthermore, as we will see later on, for a continuous scaling function with compact support linear independence implies ℓ_p-*stability*, $1 \leq p \leq \infty$, i.e., there are constants $0 < C \leq D < \infty$ such that

$$C\|u\|_{\ell_p} \leq \left\| \sum_{\beta \in \mathbb{Z}} u_\beta \varphi(\cdot - \beta) \right\|_{L_p} \leq D\|u\|_{\ell_p} \qquad (2.13)$$

holds for all $u \in \ell_p(\mathbb{Z})$. For application purposes, ℓ_p-stability is particularly important since it ensures that a small perturbation of the coefficient sequence u results only in a small perturbation of the function

$$f := \sum_{n=0}^{m-1} \sum_{\beta \in \mathbb{Z}^d} u_\beta \varphi_n(\cdot - \beta)$$

and vice versa. Thus, it is indispensable for the existance of numerically stable algorithms based on the scaling function φ. Therefore, a generalized version of the interpolation condition (2.12) plays a major role throughout this work.

2.3 Extension: Biorthogonality

As we have already stated in the introduction, within the classical orthonormal wavelet setting one is confronted with some serious limitations. For instance, if a scaling function has compact support and, in addition, is continuous, then it can neither be interpolating nor symmetric. A common strategy to bypass these restrictions is to switch to the biorthogonal setting which has been introduced in [25]. Instead of one orthonormal bases, one considers a *dual pair* of stable wavelet bases, i.e., two Riesz bases associated to the motherwavelets ψ and $\widetilde{\psi}$ via (2.1) which satisfy the *biorthogonality condition*

$$\langle \psi, 2^{j/2}\widetilde{\psi}(2^j \cdot -\beta) \rangle = \delta_{0,j}\delta_{0,\beta} \qquad (2.14)$$

for all $j, \beta \in \mathbb{Z}$. Thus, instead of analyzing a function $f \in L_2(\mathbb{R})$ by means of one orthonormal wavelet basis, one obtains two expansions of the form (2.7) in terms of ψ and $\widetilde{\psi}$, respectively. From (2.14) it is clear that the wavelet coeficients are determined by

$$d_{j,\beta} := 2^{j/2}\langle f, \widetilde{\psi}(2^j \cdot -\beta)\rangle \quad \text{and} \quad \widetilde{d}_{j,\beta} := 2^{j/2}\langle f, \psi(2^j \cdot -\beta)\rangle.$$

Therefore, given a nice function in terms of Section 2.2, the vanishing moments of the dual wavelet $\widetilde{\psi}$ determine the sparsity of the representation with respect to the primal wavelet ψ and vice versa.

Similar to the orthonormal case, ψ and $\widetilde{\psi}$ are usually constructed by means of two multiresolution analyses $(V_j)_{j\in\mathbb{Z}}$ and $(\widetilde{V}_j)_{j\in\mathbb{Z}}$, respectively. However, for the biorthogonal case, the classical definition of an MRA has to be modified slightly, i.e., the requirement (MRA5) is substituted by

(MRA5') there exists $\varphi \in L_2(\mathbb{R})$ such that $\{\varphi(x - \beta)\,|\,\beta \in \mathbb{Z}\}$ is a Riesz basis in V_0.

Consequently, the integer translates of the primal generator φ and the dual generator $\widetilde{\varphi}$, respectively, do not have to be orthonormal but ℓ_2–stable. In addition, to lead to biorthogonal wavelet bases, these scaling functions have to be *biorthogonal*, i.e., the integer translates of φ and $\widetilde{\varphi}$ are mutually orthogonal. Then the corresponding pair of MRA is called *dual*.

Given a dual pair of MRA, to construct the biorthogonal wavelet bases, one proceeds similar to the orthonormal case. First of all, for every $j \in \mathbb{Z}$ one defines some, in general not orthogonal complement spaces W_j of V_j in V_{j+1} and $\widetilde{W}_j$ of $\widetilde{V}_j$ in $\widetilde{V}_{j+1}$, respectively. These spaces have to satisfy the additional requirement

$$W_j \perp \widetilde{V}_j \quad \text{and} \quad \widetilde{W}_j \perp V_j$$

such that the two multiresolution analyses are coupled crosswise. As in the orthonormal setting, the next step is to find two mother wavelets ψ and $\widetilde{\psi}$ such that their integer translates constitute Riesz bases of W_0 and $\widetilde{W}_0$, respectively. However, in contrast to the scalar case, showing that the dilates and translates of these two mother wavelets give rise to biorthogonal wavelet bases is somewhat more involved. Nevertheless, in [25] the following analogon of Theorem 2.1.1 has been shown.

Theorem 2.3.1. *Let φ and $\widetilde{\varphi}$ be biorthogonal generators of an MRA with finitely supported masks a and $\widetilde{a}$, respectively. Furthermore, let their Fourier transforms satisfy*

$$|\widehat{\varphi}(\omega)| \le C(1 + |\omega|)^{-1/2-\varepsilon} \quad \text{and} \quad |\widehat{\widetilde{\varphi}}(\omega)| \le C(1 + |\omega|)^{-1/2-\varepsilon}$$

for some $C, \varepsilon > 0$ and almost all $\omega \in \mathbb{R}$. Then, with

$$b_\beta := (-1)^\beta \widetilde{a}_{1-\beta} \quad \text{and} \quad \widetilde{b}_\beta := (-1)^\beta a_{1-\beta},$$

the functions

$$\psi := \sum_{\beta \in \mathbb{Z}} b_\beta \varphi(2 \cdot - \beta) \qquad \text{and} \qquad \widetilde{\psi} := \sum_{\beta \in \mathbb{Z}} b_\beta \widetilde{\varphi}(2 \cdot - \beta)$$

give rise to biorthogonal wavelet bases.

Hence, also for the biorthogonal case there exist some canonical wavelets for a given pair of generators.

2.4 Stationary Subdivision

Closely related to the theory of multiresolution analyses and scaling functions is the field of stationary subdivision schemes. In general, subdivision methods are a class of recursive algorithms which have found widespread use in computer graphics for computing curves and surfaces. For a detailed discussion of this topic we refer to the survey article [14].

In the following, we give a brief introduction to the field of stationary subdivision schemes and their relation to scaling vectors. The starting point of a stationary subdivision scheme is the *subdivision operator* S_a associated to a finitely supported mask or sequence $a := (a_\beta)_{\beta \in \mathbb{Z}}$. For an arbirtrary sequence $u := (u_\beta)_{\beta \in \mathbb{Z}}$, S_a is the linear operator defined by the equation

$$(S_a u)_\alpha := \sum_{\beta \in \mathbb{Z}} u_\beta a_{\alpha - 2\beta}, \quad \alpha \in \mathbb{Z}.$$

The corresponding subdivision scheme is obtained as follows. Assume the sequence $u_0 := u$ represents a polygonal curve P_u, i.e., a piecewise linear function with nodes $\beta \in \mathbb{Z}$ and $P_u(\beta) = u_\beta$. Then, for a given mask a, the associated subdivision operator S_a defines a rule to obtain the representation $u_1 := S_a u$ of a polygonal curve on the denser set of nodes $2^{-1}\mathbb{Z}$. Consequently, iterating this procedure yields polygonal curves represented by

$$u_n := S_a u_{n-1} = S_a^n u, \qquad n > 0, \tag{2.15}$$

with nodes in $2^{-n}\mathbb{Z}$ becoming ever denser. The sequence of operators $(S_a^n)_{n \in \mathbb{Z}_+}$ is called a *stationary subdivision scheme*. The term *stationary* refers to using the same mask a in each step of the iteration (2.15). Since a subdivision scheme is determined by the corresponding subdivision operator, in the following we use both terms synonymously.

Now, it is natural to ask for the convergence of this scheme. For this reason, one often assumes that the sequence u is bounded. Thus, the subdivision scheme associated to a mask a is said to *converge* for $u \in \ell_\infty(\mathbb{Z})$ if there exists a continuous function f_u such that

$$\lim_{n \to \infty} \left\| \left(f_u(2^{-n}\beta) \right)_{\beta \in \mathbb{Z}} - S_a^n u \right\|_\infty = 0.$$

The subdivision scheme S_a is *convergent* if it converges for all $u \in \ell_\infty(\mathbb{Z})$ and there exists at least one u such that the limit function satisfies $f_u \neq 0$. Hence, starting with a polygonal curve represented by $u \in \ell_\infty(\mathbb{Z})$, a convergent subdivision scheme provides a method to generate a continuous, in general non-polygonal curve given by the limit function f_u. Note that this curve does not necessarily interpolate the original points, i.e., f_u may or may not satisfy

$$f_u(\beta) = u_\beta, \quad \beta \in \mathbb{Z}. \tag{2.16}$$

However, if for all sequences $u \in \ell_\infty(\mathbb{Z})$ the limit function satisfies (2.16), then the corresponding subdivision scheme is called *interpolatory*, cf. [92].

To study the properties of the limit function f_u, the following theorem which connects subdivision schemes with the refinability of functions is very useful. For a proof we refer to [14].

Theorem 2.4.1. *For a finitely supported mask a let the associated subdivision scheme S_a be convergent. Then there exists a unique continuous function φ with compact support which satisfies the refinement equation (2.3) with the mask a and*

$$\sum_{\beta \in \mathbb{Z}} \varphi(\cdot - \beta) = 1.$$

Moreover, for each $u \in \ell_\infty(\mathbb{Z})$ the limit function f_u satisfies

$$f_u(x) = \sum_{\beta \in \mathbb{Z}} u_\beta \varphi(x - \beta), \quad x \in \mathbb{R}.$$

If the integer translates of a refinable function φ are linearly independent or at least ℓ_∞-stable then also the converse result holds.

Theorem 2.4.2. *Let φ be a continuous compactly supported function which is refinable with respect to a finitely supported mask a. Furthermore, let the integer translates of φ be ℓ_∞-stable. Then the associated subdivision scheme S_a is convergent.*

It has been shown in [72] that for a continuous function with compact support ℓ_2-stability is equivalent to ℓ_p-stability for all $1 \leq p \leq \infty$. Consequently, since (MRA5) and (MRA5'), respectively, imply that the integer translates of a scaling function φ are at least ℓ_2-stable, we observe that a generator can not only be used to construct wavelet bases by means of an MRA, but also may give rise to a convergent stationary subdivision scheme. Moreover, if in addition φ is interpolating then the above results show that the corresponding subdivision scheme is interpolatory.

Chapter 3

Starter: The General Setting

In this chapter, we extend the concepts introduced in the preceding chapter to a more general setting. This extension is twofold. Motivated by the observation that in some cases the scalar setting is somewhat restricted, we focus on multiwavelets. This vector valued approach provides additional flexibility which can be used to construct wavelets with more or nicer properties as the scalar setting allows. Nevertheless, one still obtains orthonormal or birthogonal bases of L_2. On the other hand, we are interested in multivariate (multi-)wavelets. Therefore, we focus on the concept of generalized scalings by means of scaling matrices.

This chapter is organized as follows. First of all, we give a brief introduction to the field of refinable function vectors or scaling vectors. Similar to scaling functions, these vector valued generators can be used to construct multiwavelets by means of a generalized multiresolution analysis and, moreover, are closely related to certain subdivision schemes. Furthermore, we introduce an interpolation property for this vector setting which, as the key ingredient in all our construction methods, plays a fundamental role throughout this work. In addition, several other properties of scaling vectors which are desirable for application purposes are discussed.

Most of the concepts introduced in this chapter appear as generalizations of the corresponding terms in the scalar wavelet setting. For a detailed discussion, see the textbooks [23, 37, 118] and the doctoral thesis [104].

3.1 Interpolating Scaling Vectors

As we have seen in the preceding chapter, one of the most important tools for a systematic construction of wavelets is the scaling function. Therefore, we start this section by introducing the more general concept of scaling vectors, i.e., vectors of functions in $L_2(\mathbb{R}^d)$ which satisfy a refinement equation similar to (2.3).

From the constructional as well as from the application related point of view it is highly appreciated that scaling functions or vectors are interpolating in the sense of Equation (2.12). Therefore, we propose an interpolation condition for the vector setting which appears as a natural extension of its scalar counterpart. In addition, we discuss in detail how scaling vectors can be used to construct multiwavelets and shed some light on their relation to a certain class of subdivision schemes.

3.1.1   Refinable Function Vectors

Let $\Phi := (\phi_0, \ldots, \phi_{r-1})^\top$, $r > 0$, be a vector of $L_2(\mathbb{R}^d)$–functions which satisfies a *matrix refinement equation*

$$\Phi(x) = \sum_{\beta \in \mathbb{Z}^d} A_\beta \Phi(Mx - \beta), \quad A_\beta \in \mathbb{R}^{r \times r}, \tag{3.1}$$

with the *mask* $A := (A_\beta)_{\beta \in \mathbb{Z}^d}$ and a *scaling matrix* $M \in \mathbb{Z}^{d \times d}$, then Φ is called (A, M)–*refinable*. For conciseness, we call Φ an r–*scaling vector* if the mask and the scaling matrix are clear from the context. The scaling matrix M has to be *expanding*, i.e., all eigenvalues of M have to be larger than one in modulus, and as a shorthand notation we use $m := |\det(M)|$. For $n, k > 0$, the space of all sequences of real $n \times k$ matrices on $\mathbb{Z}^d$ is denoted by $\ell(\mathbb{Z}^d)^{n \times k}$. Thus, we have $A \in \ell(\mathbb{Z}^d)^{r \times r}$, and the mask entries are denoted by

$$A_\beta = \begin{pmatrix} a_\beta^{(0,0)} & \cdots & a_\beta^{(0,r-1)} \\ \vdots & \ddots & \vdots \\ a_\beta^{(r-1,0)} & \cdots & a_\beta^{(r-1,r-1)} \end{pmatrix}. \tag{3.2}$$

In many cases, we will consider masks consisting of a finite number of nonvanishing coefficients only, the corresponding sequence space will be denoted by $\ell_0(\mathbb{Z}^d)^{r \times r}$.

Applying the Fourier transform component-wise to (3.1) yields

$$\widehat{\Phi}(\omega) = \frac{1}{m} \mathbf{A}(e^{-iM^{-\top}\omega}) \widehat{\Phi}(M^{-\top}\omega), \quad \omega \in \mathbb{R}^d, \tag{3.3}$$

where $e^{-i\omega}$ is a shorthand notation for $(e^{-i\omega_1}, \ldots, e^{-i\omega_d})^\top$. The *symbol* $\mathbf{A}(z)$ is the matrix valued Laurent series with entries

$$a_{i,j}(z) := \sum_{\beta \in \mathbb{Z}^d} a_\beta^{(i,j)} z^\beta, \quad z \in \mathbb{T}^d,$$

and $\mathbb{T}^d := \{z \in \mathbb{C}^d : |z_i| = 1, i = 1, \ldots, d\}$ denotes the d–dimensional torus. All elements of $\mathbb{T}^d$ have the form $z = e^{-i\omega}$, $\omega \in \mathbb{R}^d$, thus we have $z^\beta = e^{-i\langle \omega, \beta \rangle}$, and for

$\xi \in \mathbb{R}^d$ we use the notation $z_\xi := e^{-i(\omega + 2\pi\xi)}$. In addition, we define $z^M := e^{-iM^\top \omega}$ such that $(z^M)^\beta = z^{M\beta}$ and $z_\xi^M := e^{-iM^\top(\omega + 2\pi\xi)}$.

The transformed refinement equation (3.3) is one of the most important tools for the construction of scaling vectors. On one hand, it directly implies $m\widehat{\Phi}(0) = \mathbf{A}(1)\widehat{\Phi}(0)$, where $\mathbf{1} := (1, \ldots, 1)^\top \in \mathbb{C}^d$. Thus, either $\widehat{\Phi}(0)$ is an m–eigenvector of $\mathbf{A}(1)$ or we have $\widehat{\Phi}(0) = 0$ which is rather undesirable as we will see in the sequel. On the other hand, iterating (3.3) yields

$$\widehat{\Phi}(\omega) = \prod_{j=1}^\infty \frac{1}{m}\mathbf{A}(e^{-iM^{-j^\top}\omega})\widehat{\Phi}(0) =: P(\omega)\widehat{\Phi}(0),$$

where, of course, convergence has to be clarified. However, if the infinite product $P(\omega)$ converges then the scaling vector Φ is completely determined by its symbol or mask, respectively, up to a constant. The following theorem, stated in [11], puts the above observations on a sound mathematical foundation and provides us with a sufficient condition for the existence of a compactly supported solution of the refinement equation (3.1), see also [77].

Theorem 3.1.1. *For a mask $A \in \ell_0(\mathbb{Z}^d)^{r \times r}$ let $\mathbf{A}(1)$ have the eigenvalues $\lambda_1 = m$, $|\lambda_2|, \ldots, |\lambda_r| < m$, then the following statements hold:*

(i) *The infinite matrix product $P(\omega)$ converges uniformly on compact sets.*

(ii) *Any m–eigenvector v of $\mathbf{A}(1)$ defines a compactly supported distributional solution Φ of (3.1) via $\widehat{\Phi}(\omega) := P(\omega)v$.*

(iii) *If Φ is a nontrivial compactly supported distributional solution of (3.1) then $\widehat{\Phi}(0)$ is an m–eigenvector of $\mathbf{A}(1)$.*

Since compact support is crucial for almost all application purposes, we focus on compactly supported scaling vectors and therefore on masks belonging to $\ell_0(\mathbb{Z}^d)^{r \times r}$.

Occasionally, we have to decompose a symbol $\mathbf{A}(z)$ into its subsymbols as follows. For a scaling matrix M let $R := \{\rho_0, \ldots, \rho_{m-1}\}$ denote a complete set of representatives of $\mathbb{Z}^d / M\mathbb{Z}^d$, and for $\rho \in R$ let $[\rho]$ denote the corresponding coset. Thus, $\mathbb{Z}^d$ decomposes into the disjoint union

$$\mathbb{Z}^d = \bigcup_{\rho \in R}[\rho] = \bigcup_{\rho \in R}(M\mathbb{Z}^d + \rho),$$

cf. Prop. 5.5 in [118]. The *i*th *subsymbol* $\mathbf{A}_i(z)$ of $\mathbf{A}(z)$ is defined by

$$\mathbf{A}_i(z) := \sum_{\beta \in \mathbb{Z}^d} A_{M\beta + \rho_i} z^\beta, \quad 0 \le i < m, \tag{3.4}$$

such that we have the decomposition

$$\mathbf{A}(z) = \sum_{i=0}^{m} \sum_{\beta \in \mathbb{Z}^d} A_{M\beta + \rho_i} z^{M\beta + \rho_i} = \sum_{i=0}^{m} z^{\rho_i} \mathbf{A}_i(z^M). \tag{3.5}$$

We close this section with some additional notation. A complete set of representatives of $\mathbb{Z}^d / M^\top \mathbb{Z}^d$ shall be denoted by $\widetilde{R} := \{\tilde{\rho}_0, \dots, \tilde{\rho}_{m-1}\}$. Furthermore, for the characteristic functions of the cosets $[\rho]$ (and $[\tilde{\rho}]$ respectively) we use the notation $\mathbb{1}_{[\rho]}(\cdot)$ (and $\mathbb{1}_{[\tilde{\rho}]}(\cdot)$). The sets R and $\widetilde{R}$ are connected by the following lemma on character sums, cf. [22].

Lemma 3.1.2. *For $\rho_i, \rho_j \in R$ and $z \in \mathbb{T}^d$ it holds that*

$$\sum_{\tilde{\rho} \in \widetilde{R}} z_{M^{-\top}\tilde{\rho}}^{\rho_i} z_{M^{-\top}\tilde{\rho}}^{-\rho_j} = m \cdot \delta_{i,j}.$$

3.1.2 The Interpolation Property

One central aim of this work is the construction of families of *interpolating m–scaling vectors* Φ with compact support, i.e., all components of Φ are at least continuous and satisfy

$$\phi_n\left(M^{-1}\beta\right) = \delta_{\rho_n, \beta} \quad \text{for all} \quad \beta \in \mathbb{Z}^d, \, 0 \leq n < m. \tag{3.6}$$

Note that, in contrast to the scalar case, the interpolation condition (and the length of the scaling vector) is determined by the determinant of the scaling matrix. Nevertheless, it can be considered as a natural generalization of the scalar interpolation condition (2.12) as follows. First of all, the scalar interpolation condition can be extended to the multivariate setting by simply allowing $\beta \in \mathbb{Z}^d$ in Equation (2.12). Now, let $\varphi \in L_2(\mathbb{R}^d)$ be an interpolating scaling function, then the rule

$$\Phi(x) := \left(\varphi(Mx - \rho_0), \dots, \varphi(Mx - \rho_{m-1})\right)^\top \tag{3.7}$$

defines an m–scaling vector which satisfies the interpolation condition (3.6).

One advantage of interpolating scaling vectors is that they give rise to a Shannon–like sampling theorem as follows. For a compactly supported function vector $\Phi \in L_2(\mathbb{R}^d)^m$ let us define the shift-invariant space

$$S(\Phi) := \left\{ \sum_{\beta \in \mathbb{Z}^d} u_\beta \Phi(\cdot - \beta) \, \Big| \, u \in \ell(\mathbb{Z}^d)^{1 \times m} \right\}.$$

Since Φ has compact support, only a finite number of $\Phi(\cdot - \beta)$ overlap. This ensures that for an arbitrary sequence $u \in \ell(\mathbb{Z}^d)^{1 \times m}$ the above sum is locally finite, and thus $S(\Phi)$ is well defined. A direct computation shows that, if Φ is a compactly supported interpolating m–scaling vector, then for all $f \in S(\Phi)$ the representation

$$f(x) = \sum_{\beta \in \mathbb{Z}^d} \sum_{i=0}^{m-1} f\left(\beta + M^{-1}\rho_i\right) \phi_i(x - \beta) \tag{3.8}$$

holds.

Similar to the scalar case, cf. Section 2.2, an immediate consequence of the interpolation property is (algebraically) *linearly independent integer translates*, i.e., the mapping

$$u \in \ell(\mathbb{Z}^d)^{1 \times m} \mapsto \sum_{\beta \in \mathbb{Z}^d} u_\beta \Phi(\cdot - \beta)$$

is injective whenever Φ has compact support and satisfies (3.6). Furthermore, it has been shown in [72] that for a continuous m–scaling vector with compact support linear independence implies ℓ_p–stability. For ease of notation, we state the definition for the case $p = 2$ only, the general case is defined analogously, cf. (2.13). A scaling vector is said to have ℓ_2–*stable integer translates* if there exist constants $0 < C \leq D < \infty$ such that

$$C \sum_{n=0}^{m-1} \|u^{(n)}\|_{\ell_2}^2 \leq \left\| \sum_{n=0}^{m-1} \sum_{\beta \in \mathbb{Z}^d} u_\beta^{(n)} \phi_n(\cdot - \beta) \right\|_{L_2}^2 \leq D \sum_{n=0}^{m-1} \|u^{(n)}\|_{\ell_2}^2 \tag{3.9}$$

holds for all $u^{(0)}, \ldots, u^{(m-1)} \in \ell_2(\mathbb{Z}^d)$. As in the scalar case, the stability of a scaling vector Φ is particularly important for the existence of numerically stable algorithms based on Φ. Moreover, it has been shown in [75] that the stability of a compactly supported scaling vector Φ has a strong impact on the properties of the symbol $\mathbf{A}(z)$ and on the properties of $S(\Phi)$ as follows.

Proposition 3.1.3. *Let Φ be a compactly supported ℓ_2–stable scaling vector with mask $A \in \ell_0(\mathbb{Z}^d)^{r \times r}$. Then the following statements hold:*

(i) *$\mathbf{A}(1)$ has a simple eigenvalue m and all other eigenvalues are smaller than m in modulus.*

(ii) *$S(\Phi)$ contains all constant functions.*

Remark 3.1.4. *The reader should observe that Proposition 3.1.3 together with Theorem 3.1.1 imply that a compactly supported ℓ_2–stable scaling vector Φ with finite mask satisfies $\widehat{\Phi}(0) \neq 0$.*

Another advantage of interpolating scaling vectors is the simple structure of their masks which considerably reduces the effort concerning their construction.

Lemma 3.1.5. *Let $\rho_k \in M\mathbb{Z}^d$, then the mask of an interpolating m–scaling vector has to satisfy*

$$a^{(i,k)}_{M\alpha+\rho_j-M^{-1}\rho_k} = \delta_{0,\alpha}\delta_{i,j} \quad \text{for all } \alpha \in \mathbb{Z}^d,\ 0 \le i,j < m.$$

Proof. For each $\gamma \in \mathbb{Z}^d$ there exists an $\alpha \in \mathbb{Z}^d$ and $j \in \{0,\ldots,m-1\}$ such that $\gamma = M\alpha + \rho_j$. Thus, we have

$$\phi_i(M^{-1}\gamma) = \phi_i(M^{-1}\rho_j + \alpha) = \sum_{\beta \in \mathbb{Z}^d} \left(a^{(i,0)}_\beta,\ldots,a^{(i,m-1)}_\beta\right)\Phi(\rho_j + M\alpha - \beta).$$

Since we have assumed that $\rho_k \in M\mathbb{Z}^d$, the interpolation condition (3.6) yields $\Phi(\beta) = \delta_{\beta,M^{-1}\rho_k}e_k$ for all $\beta \in \mathbb{Z}^d$, where e_k denotes the kth unit vector. Therefore, we obtain

$$\phi_i(M^{-1}\gamma) = a^{(i,k)}_{M\alpha+\rho_j-M^{-1}\rho_k}.$$

On the other hand, (3.6) implies $\phi_i(M^{-1}\gamma) = \delta_{0,\alpha}\delta_{i,j}$ which completes the proof. $\qquad\square$

For simplicity of notation, we shall assume $\rho_0 = 0 \in \mathbb{Z}^d$ without loss of generality. Then the above lemma implies that the symbol of an interpolating m–scaling vector has to have the form

$$\mathbf{A}(z) = \begin{pmatrix} z^{\rho_0} & a^{(0,1)}(z) & \cdots & a^{(0,m-1)}(z) \\ \vdots & \vdots & \ddots & \vdots \\ z^{\rho_{m-1}} & a^{(m-1,1)}(z) & \cdots & a^{(m-1,m-1)}(z) \end{pmatrix}. \tag{3.10}$$

For the case $m = 2$ we can choose $R = \{0, \rho\}$ and obtain

$$\mathbf{A}(z) = \begin{pmatrix} 1 & a^{(0)}(z) \\ z^\rho & a^{(1)}(z) \end{pmatrix}. \tag{3.11}$$

3.1.3   Multiwavelets

Similar to scaling functions, a common motivation for dealing with r–scaling vectors is the construction of r–*multiwavelets*. These appear as a collection of function vectors $\Psi^{(n)} := \left(\psi_0^{(n)},\ldots,\psi_{r-1}^{(n)}\right)^T \in L_2(\mathbb{R}^d)^r$, $0 < n < m$, for which

$$\left\{ \psi_0^{(n)}\left(M^j \cdot -\beta\right),\ldots,\psi_{r-1}^{(n)}\left(M^j \cdot -\beta\right) \;\middle|\; j \in \mathbb{Z}, \beta \in \mathbb{Z}^d, 0 < n < m \right\}$$

forms a (Riesz) basis of $L_2(\mathbb{R}^d)$. Compared to the classical setting introduced in Section 2.1, this notion of wavelets is much more general. First of all, at least for the moment, we drop the very restrictive orthonormal basis property and focus on stable bases only. However, in the sequel we will come back to stronger basis properties. Moreover, the above basis is not spanned by the dilates and translates of one single wavelet only, but by dilated and translated versions of several functions. This has two reasons. First of all, within the multiwavelet concept one assumes that all component functions of a multiwavelet conjointly constitute the corresponding multiwavelet basis. On the other hand, similar to the classical case, a systematical approach to the construction of multiwavelets is given by means of a multiresolution analysis. Then, the number of wavelets associated to an MRA is determined by the determinant of the scaling matrix M as above, see [90] and [24] for details.

However, the classical definition of an MRA given by Mallat and Meyer in [87, 91], cf. Section 2.1, has to be adapted to our more general setting first. Of course, this topic has been extensively treated in the literature. For example, MRA with non-orthogonal scaling functions have been introduced in [25], a generalization of the MRA concept using scaling matrices can be found in [24, 53, 90], and the vector case has been extensively studied in [40, 50], see also [74]. Combining these different approaches, we obtain the following definition. A *multiresolution analysis* is a sequence $(V_j)_{j \in \mathbb{Z}}$ of closed subspaces of $L_2(\mathbb{R}^d)$ which satisfies:

(MRA1) $V_j \subset V_{j+1}$ for each $j \in \mathbb{Z}$,

(MRA2) $g(x) \in V_j$ if and only if $g(Mx) \in V_{j+1}$ for each $j \in \mathbb{Z}$,

(MRA3) $\bigcap_{j \in \mathbb{Z}} V_j = \{0\}$,

(MRA4) $\bigcup_{j \in \mathbb{Z}} V_j$ is dense in $L_2(\mathbb{R}^d)$, and

(MRA5) there exists an ℓ_2–stable $\Phi \in L_2(\mathbb{R}^d)^r$ such that

$$V_0 = \overline{\text{span}\{\phi_i(x - \beta) \mid \beta \in \mathbb{Z}^d, 0 \leq i < r\}}.$$

Now, to obtain some multiwavelets, we first have to find a way to construct a suitable MRA. As in the scalar case, (MRA1) and (MRA2) imply that the function vector Φ in (MRA5) satisfies a refinement equation of the form (3.1). Hence, we have to study under which circumstances a scaling vector generates an MRA. Let $\Phi \in L_2(\mathbb{R}^d)^r$ be a stable scaling vector with mask $A \in \ell_0(\mathbb{Z}^d)$ (or, at least, $A \in \ell_2(\mathbb{Z}^d)^{r \times r}$, i.e., the component sequences of A are square summable). If we define V_0 via (MRA5) and V_j, $j \in \mathbb{Z}$, via (MRA2), then the refinement equation implies (MRA1). Furthermore, it has been shown in [74] that in this case also the intersection in (MRA3) is trivial. Hence, condition (MRA4) remains to be checked. The following theorem provides us with an equivalent condition in terms of $\widehat{\Phi}$, for the proof see Theorem 2.1 and Remark 2.6 in [74].

Theorem 3.1.6. *Let $\Phi \in L_2(\mathbb{R}^d)^r$ be a stable scaling vector, and define V_j, $j \in \mathbb{Z}$, via (MRA5) and (MRA2). Furthermore, assume that (MRA1) holds. Then (MRA4) is satisfied if and only if*

$$\widehat{Z}(\Phi) := \bigcap_{i=0}^{r-1} \bigcap_{j \in \mathbb{Z}} M^j \{\omega \in \mathbb{R}^d \mid \widehat{\phi}_i(\omega) = 0\}$$

is a set of measure zero.

We immediately obtain the following corollary.

Corollary 3.1.7. *Any stable compactly supported scaling vector Φ with finite mask generates an MRA via (MRA5) and (MRA2).*

Proof. As stated above, only (MRA4) has to be checked. We show that for a stable compactly supported scaling vector Φ with finite mask $\widehat{Z}(\Phi) = \emptyset$ holds. Assume there exists an $\omega_0 \in \widehat{Z}(\Phi)$, then we have $\widehat{\phi}_i(M^{-j}\omega_0) = 0$ for all $j \in \mathbb{Z}$ and $0 \leq i < r$. Since M is expanding, the spectral radius $\mathrm{spr}(M^{-1}) < 1$ and therefore $M^{-j}\omega_0 \to 0$ as $j \to \infty$. Thus, each open set containing 0 also contains a zero of $\widehat{\Phi}$. On the other hand, if Φ is compactly supported then, due to the Riemann–Lebesgue lemma, $\widehat{\Phi}$ is continuous. Furthermore, due to Remark 3.1.4, we have $\widehat{\phi}_i(0) \neq 0$ for some $0 \leq i < r$. Therefore, $\widehat{\phi}_i$ is also nonzero in a small neighborhood of 0 which contradicts the assumption $\widehat{Z}(\Phi) \neq \emptyset$. Thus, Theorem 3.1.6 implies that (MRA4) holds. □

In the sequel, we sketch how to construct multiwavelets given an MRA generated by an r–scaling vector Φ. For a detailed discussion of this construction process, see, e.g., [12, 36, 51, 87]. As in the univariate scalar case, let W_0 denote an algebraic complement of V_0 in V_1 and define $W_j := \{g(M^j \cdot) \mid g \in W_0\}$. Then, one immediately obtains that $V_{j+1} = V_j \oplus W_j$ and consequently, due to (MRA3) and (MRA4), $L_2(\mathbb{R}^d) = \bigoplus_{j \in \mathbb{Z}^d} W_j$. If one finds function vectors $\Psi^{(n)} \in L_2(\mathbb{R}^d)^r$, $0 < n < m$, such that the integer translates of the components of all $\Psi^{(n)}$ are a stable basis of W_0, then, by dilation, one obtains a stable multiwavelet basis of $L_2(\mathbb{R}^d)$. Since $W_0 \subset V_1$, each $\Psi^{(n)}$ can be represented as

$$\Psi^{(n)}(x) = \sum_{\beta \in \mathbb{Z}^d} B_\beta^{(n)} \Phi(Mx - \beta) \tag{3.12}$$

for some $B^{(n)} \in \ell(\mathbb{Z}^d)^{r \times r}$. By applying the Fourier transform component wise to (3.12) one obtains

$$\widehat{\Psi}^{(n)}(\omega) = \frac{1}{m} \mathbf{B}^{(n)}(e^{-iM^{-\top}\omega}) \widehat{\Phi}(M^{-\top}\omega), \quad \omega \in \mathbb{R}^d,$$

with the symbol

$$\mathbf{B}^{(n)}(z) := \sum_{\beta \in \mathbb{Z}^d} B_\beta^{(n)} z^\beta, \quad z \in \mathbb{T}^d.$$

Therefore, the task of finding a stable multiwavelet basis can be reduced to constructing the symbols $\mathbf{B}^{(n)}(z)$. It should be mentioned that, in general, the choice of the complement space W_0 is by no means unique. Hence, there may be several multiwavelet bases associated to one scaling vector. On the other hand, it is still an open problem wether there always exists a stable multiwavelet bases associated to a given MRA or not, cf. [12].

3.1.4 Vector Subdivision

Similar to the univariate scalar case, also to a mask $A \in \ell_0(\mathbb{Z}^d)^{r \times r}$ a *subdivision operator* $\mathcal{S}_A$ can be associated, i.e., for a sequence in $u \in \ell(\mathbb{Z}^d)^{1 \times r}$ we define

$$(\mathcal{S}_A u)_\alpha := \sum_{\beta \in \mathbb{Z}^d} u_\beta A_{\alpha - M\beta}, \quad \alpha \in \mathbb{Z}^d.$$

Again, the rule

$$u_0 := u \quad \text{and} \quad u_n := \mathcal{S}_A u_{n-1} = \mathcal{S}_A^n u, \quad n > 0,$$

defines a stationary *vector subdivision scheme*.

For the convergence analysis of this scheme, one often considers the space of $1 \times r$–vector valued sequences with bounded component sequences denoted by $\ell_\infty(\mathbb{Z}^d)^{1 \times r}$. The norm $\|.\|_\infty$ on this space is obtained as the usual product norm, i.e., for $u := (u_\beta^{(0)}, \ldots, u_\beta^{(r-1)})_{\beta \in \mathbb{Z}^d} \in \ell_\infty(\mathbb{Z}^d)^{1 \times r}$ one has $\|u\|_\infty := \sup_{\beta \in \mathbb{Z}^d} \max_{0 \leq i < r} |u_\beta^{(i)}|$. A vector subdivision scheme is said to *converge* for $u \in \ell_\infty(\mathbb{Z}^d)^{1 \times r}$ if there exists a continuous vector valued function $\mathbf{f}_u := (f_u^{(0)}, \ldots, f_u^{(r-1)})$ such that

$$\lim_{n \to \infty} \|(\mathbf{f}_u(M^{-n}\beta))_{\beta \in \mathbb{Z}^d} - \mathcal{S}_A^n u\|_\infty = 0.$$

As in the classical case, a vector subdivision scheme is called *convergent* if it converges for all $u \in \ell_\infty(\mathbb{Z}^d)^{1 \times r}$ and there exists at least one u such that $\mathbf{f}_u \neq 0$.

For $r = 1$ the interpretation of this scheme is pretty similar to the classical case discussed in Section 2.4. One assumes that the sequence $u \in \ell_\infty(\mathbb{R}^d)$ represents a polyhedral hypersurface with nodes in $\mathbb{Z}^d$. Then the subdivision scheme generates a sequence of polyhedral hypersurfaces with nodes in $M^{-n}\mathbb{Z}^d$ which converge to some limiting hypersurface determined by the scalar valued function $\mathbf{f}_u$. Thus, by substituting the scaling parameter 2 by M and allowing $\beta \in \mathbb{Z}^d$ in Section 2.4, all results of the univariate case are carried forward to the multivariate case, cf. [14, 34].

For the vector case, the corresponding results are somewhat more involved. To establish the relation between vector subdivision schemes and scaling vectors, we first have to consider a slightly more general setting. Obviously, the subdivision operator $\mathcal{S}_A$ can be applied to matrix valued sequences $U \in \ell(\mathbb{Z}^d)^{r \times r}$ as well by interpreting each row of U as an input sequence for $\mathcal{S}_A$. Hence, if a vector subdivision scheme is convergent, then it converges for a matrix valued sequence with rows in $\ell_\infty(\mathbb{Z}^d)^{1 \times r}$ to a matrix valued limit function. Let $\mathbf{F}_\delta$ denote the so-called *canonical limit function* corresponding to the sequence $(\delta_{0,\beta} \mathbf{I}_r)_{\beta \in \mathbb{Z}^d}$. It has been shown in [34], see also [105], that for each $u \in \ell_\infty(\mathbb{Z}^d)^{1 \times r}$ the limit function $\mathbf{f}_u$ is determined by the canonical limit function as follows.

Theorem 3.1.8. *Let $\mathbf{F}_\delta$ be the canonical limit function of a convergent stationary vector subdivision scheme associated to a mask $A \in \ell_0(\mathbb{Z}^d)^{r \times r}$. Then for each $u \in \ell_\infty(\mathbb{Z}^d)^{1 \times r}$ the limit function $\mathbf{f}_u$ satisfies*

$$\mathbf{f}_u = \sum_{\beta \in \mathbb{Z}^d} u_\beta \mathbf{F}_\delta(\cdot - \beta).$$

In addition, the columns of $\mathbf{F}_\delta$ satisfy the refinement equation (3.1).

Hence, if the vector subdivision scheme converges and there exists a solution Φ of the refinement equation (3.1) which is unique up to multiplication with a constant, then $\mathbf{F}_\delta$ has rank 1. Consequently, the corresponding vector subdivision scheme is called a *rank 1 subdivision scheme*. The following converse result was obtained in [34] as well.

Theorem 3.1.9. *Let Φ be a continuous scaling vector with compact support that corresponds to a mask $A \in \ell_0(\mathbb{Z}^d)^{r \times r}$. Furthermore, let the integer translates of Φ be linearly independent. Then, the stationary vector subdivision scheme induced by the subdivision operator $\mathcal{S}_A$ associated to A is convergent and, moreover, of rank 1.*

Remark 3.1.10. *The convergence of subdivision schemes can also be studied in an ℓ_p-sense for $1 \le p \le \infty$. Then analog results can be obtained where the notion of linear independence is substituted by ℓ_p-stability. For ease of notation, we omit a discussion of this more general setting and refer to [34]. Moreover, since we focus on interpolating scaling vector which consequently possess linearly independent integer translates and are continuous as well, the above setting does perfectly fit.*

In contrast to the scalar case, vector subdivision schemes call for a rather more sophisticated interpretation. One possibility is to assume that the vector valued starting sequence $u \in \ell_\infty(\mathbb{Z}^d)^{1 \times r}$ represents r polyhedral hypersurfaces which lead to r distinct hypersurfaces given by the limit function $\mathbf{f}_u$. However, as we will see in the sequel, this interpretation is not suitable for rank 1 subdivision schemes. Let $A \in \ell_0(\mathbb{Z}^d)^{r \times r}$ be a mask which leads to a convergent rank 1 subdivision scheme. Furthermore, let Φ be a nontrivial column of the corresponding canonical limit function $\mathbf{F}_\delta$. Hence, there exists a vector $y \in \mathbb{R}^r$ such that $\mathbf{F}_\delta = \Phi y^\top$. It immediately follows from Theorem 3.1.8 that for each $u \in \ell_\infty(\mathbb{Z}^d)^{1 \times r}$ the limit function $\mathbf{f}_u$ satisfies $\mathbf{f}_u(x) = f_u(x) y^\top$, where f_u is given by

$$f_u := \sum_{\beta \in \mathbb{Z}^d} u_\beta \Phi(\cdot - \beta).$$

Thus, although the components of u may be independently chosen, all hypersurfaces given by $\mathbf{f}_u$ coincide up to multiplication with a constant. Quite recently, it has been proposed in [20, 21] that only one component of u, say $u^{(0)}$, represents a polyhedral hypersurface. The other components $u^{(1)}, \ldots, u^{(r-1)}$ are considered to be parameters which form the shape of the limit hypersurface which is given by the component $f_u^{(0)}$ of the limit function $\mathbf{f}_u$. With this interpretation it is quite natural to ask wether the subdivision scheme is *interpolatory* in the sense that

$$f_u^{(0)}(\beta) = u_\beta^{(0)},$$

i.e., the limit hypersurface contains the vertices of the starting polyhedral hypersurface. In [20] it has been shown that under certain mild conditions a stationary rank 1 vector subdivision scheme associated to a mask A is interpolatory if A satisfies

$$a_{M\beta}^{(i,0)} = \delta_{i,0} \delta_{0,\beta}.$$

Hence, due to Lemma 3.1.5, we observe that an interpolating scaling vector leads to an interpolatory rank 1 vector subdivision scheme. On the other hand, interpolatory subdivision schemes in the above sense do only preserve one component of the input sequence u. In [28] another approach has been proposed in which all components of u are preserved, i.e., for each $i \in \{0, \ldots, r-1\}$ there exists an $\iota = \iota(i) \in \{0, \ldots, r-1\}$ and an $\alpha = \alpha(i) \in \mathbb{Z}^d$ such that for all $u \in \ell_\infty(\mathbb{Z}^d)^{1 \times r}$

$$(\mathcal{S}_A u)_{M\beta + \alpha}^{(\iota)} = u_\beta^{(i)}.$$

Thus, the components of u appear as subsequences of some components of $\mathcal{S}_A u$. Then, the corresponding subdivision scheme is called *data preserving*. A necessary and sufficient condition is that

$$a_{M\beta + \alpha}^{(j,\iota)} = \delta_{0,\beta} \delta_{j,i}, \qquad j = 0, \ldots, r-1,$$

holds true for all $\beta \in \mathbb{Z}^d$. Again, Lemma 3.1.5 yields that interpolating scaling vectors do also lead to data preserving subdivision schemes with $\alpha(i) = \rho_i$ and $\iota(i) = 0$. Moreover, the approach in [28] completely resembles our notion of interpolating scaling vectors.

3.1.5   The Transition Operator

In addition to the subdivision operator, to each mask $A \in \ell_0(\mathbb{Z}^d)^{r \times r}$ another linear operator can be associated which is closely related to the properties of the corresponding scaling vector. For a sequence in $\ell_0(\mathbb{Z}^d)^r$ we define the *transition operator* $\mathcal{T}_A$ by

$$(\mathcal{T}_A v)_\alpha := \sum_{\beta \in \mathbb{Z}^d} A_{M\alpha - \beta} v_\beta, \quad \alpha \in \mathbb{Z}^d, \, v \in \ell_0(\mathbb{Z}^d)^r.$$

If we introduce the bilinear form

$$\langle u, v \rangle := \sum_{\alpha \in \mathbb{Z}^d} u_{-\alpha} v_\alpha, \quad u \in \ell^{1 \times r}(\mathbb{Z}^d), \, v \in \ell_0^r(\mathbb{Z}^d),$$

then we immediately obtain

$$\langle \mathcal{S}_A u, v \rangle = \sum_{\alpha \in \mathbb{Z}^d} \sum_{\beta \in \mathbb{Z}^d} u_\beta A_{-\alpha - M\beta} v_\alpha = \langle u, \mathcal{T}_A v \rangle.$$

Consequently, the transition operator $\mathcal{T}_A$ is the algebraic adjoint of the subdivision operator $\mathcal{S}_A$.

In the literature, the term transition operator has an ambiguous meaning. It is also used for the operator $\mathbf{T_A}$, associated to the symbol $\mathbf{A}(z)$, defined by

$$\mathbf{T_A} \mathbf{C}(z) := \frac{1}{m^2} \sum_{\tilde{\rho} \in \tilde{R}} \mathbf{A}\left(z_{\tilde{\rho}}^{M^{-1}}\right) \mathbf{C}\left(z_{\tilde{\rho}}^{M^{-1}}\right) \overline{\mathbf{A}\left(z_{\tilde{\rho}}^{M^{-1}}\right)}^{\top}, \quad z \in \mathbb{T}^d,$$

for all $r \times r$ matrices of Laurent polynomials $\mathbf{C}(z)$. It was shown in [75] that $\mathbf{T_A}$ is a linear operator which, restricted to a certain finite dimensional space $\mathbb{H}$ which is invariant under $\mathbf{T_A}$, has the matrix representation

$$\mathbf{T_A}\big|_{\mathbb{H}} \quad \sim \quad (\mathfrak{A}_{M\alpha - \beta})_{\alpha, \beta \in K}.$$

Here, $\mathfrak{A} \in \ell_0(\mathbb{Z}^d)^{r^2 \times r^2}$ is the mask defined by

$$\mathfrak{A}_\alpha := \frac{1}{m} \sum_{\beta \in \mathbb{Z}^d} A_{\beta - \alpha} \otimes \overline{A_\beta}, \quad \alpha \in \mathbb{Z}^d,$$

and $\otimes$ denotes the *Kronecker product*, i.e., $P \otimes Q := (p_{i,j}Q)_{i,j}$ for arbitrary matrices P and Q. The index set K is given by

$$K := \left(\sum_{n=1}^{\infty} M^{-n} \operatorname{supp}(\mathfrak{A}) \right) \cap \mathbb{Z}^d \qquad (3.13)$$

with $\operatorname{supp}(\mathfrak{A}) := \{\beta \in \mathbb{Z}^d \,|\, \mathfrak{A}_\beta \neq 0\}$. Thus, $\mathbf{T_A}|_\mathbb{H}$ corresponds to a truncation of the matrix representation of the transition operator $\mathcal{T}_\mathfrak{A}$ with respect to the mask $\mathfrak{A} \in \ell_0(\mathbb{Z}^d)^{r^2 \times r^2}$. The corresponding restricted transition operator will be denoted by $\mathcal{T}_{\mathfrak{A},K}$. Furthermore, it was shown in [75], see also [71], that all eigenfunctions of $\mathbf{T_A}$ corresponding to nonzero eigenvalues belong to $\mathbb{H}$. Hence, we have

$$\operatorname{spec}(\mathbf{T_A}) \setminus \{0\} = \operatorname{spec}(\mathcal{T}_{\mathfrak{A},K}) \setminus \{0\},$$

where spec denotes the spectrum of an operator or a matrix. These spectra play a crucial role in studying the stability and regularity of a scaling vector. We will come back to this topic in Sections 3.2.1 and 3.2.3, respectively.

Another feature of the transition operator is that it can be used to compute the function values of a scaling vector on the integers. Let Φ be a continuous scaling vector with compact support. The refinement equation (3.1) implies that the sequence $(\Phi(\beta))_{\beta \in \mathbb{Z}^d}$ is an 1–eigenvector of $\mathcal{T}_A$. Hence, if the eigenvalue 1 of $\mathcal{T}_A$ is simple, a corresponding eigenvector determines $(\Phi(\beta))_{\beta \in \mathbb{Z}^d}$ up to multiplication with a constant. Since Φ is compactly supported, we only have to consider the truncated transition operator $\mathcal{T}_{A,K'}$, and it has been shown in [61, 75] that the index set K' is determined by

$$K' := \left(\sum_{n=1}^{\infty} M^{-n} \operatorname{supp}(A) \right) \cap \mathbb{Z}^d,$$

compare with (3.13). The main problem in implementing this method is the computation of the set K', since it is defined by an infinite Minkowski sum. This obstacle has been overcome in [58], where an iterative algorithm for computing K' is given. Of course, this algorithm can be used for computing the set K in (3.13) as well.

3.2   Desirable Properties

Though our main aim is to construct interpolating scaling vectors with compact support and some corresponding multiwavelet bases, we also intend to incorporate several additional properties. First of all, the multiwavelets obtained by our approach ought to constitute *nice* $L_2(\mathbb{R}^d)$–bases, preferably orthonormal or biorthogonal bases. In addition, as we have seen in Section 2.2, for application purposes the multiwavelets are desired to possess a high order of vanishing moments as well as a certain smoothness. These properties are also needed for the characterization of several function spaces, see, e.g., [46, 85] for details. On the other hand, most properties of a multiwavelet are inherited from the underlying scaling vector, e.g., the minimum number of vanishing moments of a multiwavelet is determined by the approximation properties of the underlying scaling vector.

For the reader's convenience we recall the basic definitions and give a short discussion of these properties in this section. In particular, we place a special focus on the characteristics of scaling vectors which lead to the desired properties of the corresponding multiwavelets.

3.2.1   Nice Basis Properties

We have seen in Section 3.1.3 that a compactly supported r–scaling vector $\Phi = (\phi_0, \ldots, \phi_{r-1})^\top$ with ℓ_2–stable integer translates as in (3.9) generates a multiresolution analysis. As a consequence, Φ may be used to construct a stable multiwavelet basis. To generate wavelet bases with stronger properties, the scaling vector has to satisfy some additional conditions.

As in the classical case, the most appreciated form of a multiwavelet basis is an orthonormal basis. Nevertheless, to obtain some additional constructional flexibility it is often beneficial to restrict oneself to the slightly weaker concept of biorthogonal multiwavelet bases. Assume we have two multiwavelets $\Psi^{(n)}$ and $\widetilde{\Psi}^{(n)}$ which give rise to stable multiwavelet bases. Furthermore, assume that their component functions satisfy

$$\left\langle \psi_i^{(n)}, m^{j/2}\widetilde{\psi}_{i'}^{(n')}(M^j \cdot -\beta) \right\rangle = c \cdot \delta_{i,i'}\delta_{n,n'}\delta_{0,j}\delta_{0,\beta} \tag{3.14}$$

for $0 \le i, i' < r$, $1 \le n, n' < m$, $j \in \mathbb{Z}$, $\beta \in \mathbb{Z}^d$, and a constant $c > 0$. Then $\Psi^{(n)}$ and $\widetilde{\Psi}^{(n)}$ are called *biorthogonal multiwavelets*, and the corresponding multiwavelet bases are called *biorthogonal*. Hence, orthonormal multiwavelet bases appear as a special case of this concept via the conditions $c = 1$ and $\Psi^{(n)} = \widetilde{\Psi}^{(n)}$ for $1 \le n < m$. The main benefit of biorthogonal multiwavelet bases is that for all $f \in L_2(\mathbb{R}^d)$ the coefficients $d_{j,\beta,n}$ of the multiwavelet expansion

$$f = \sum_{j=-\infty}^{\infty} \sum_{\beta \in \mathbb{Z}^d} \sum_{n=1}^{m-1} d_{j,\beta,n}^\top \Psi^{(n)}(M^j \cdot -\beta) \tag{3.15}$$

are given by the inner products

$$d_{j,\beta,n} = \frac{m^j}{c}\langle f, \widetilde{\Psi}^{(n)}(M^j \cdot -\beta)\rangle, \tag{3.16}$$

where $\langle f, \widetilde{\Psi} \rangle := (\langle f, \widetilde{\psi}_0 \rangle, \dots, \langle f, \widetilde{\psi}_{r-1} \rangle)^\top$. Thus, in practice, the expansion of $f \in L_2(\mathbb{R}^d)$ can actually be computed or at least be approximated.

Usually, the starting point for the construction of biorthogonal multiwavelet bases are biorthogonal scaling vectors. Two r–scaling vectors Φ and $\widetilde{\Phi}$ are called *biorthogonal* or *duals* of each other if the integer translates of all component functions are mutually orthogonal, i.e.,

$$\left\langle \phi_i, \widetilde{\phi}_j(\cdot - \beta) \right\rangle = c \cdot \delta_{i,j} \delta_{0,\beta}, \quad 0 \leq i, j < r, \tag{3.17}$$

holds for all $\beta \in \mathbb{Z}^d$ and a constant $c > 0$. If we have $\Phi = \widetilde{\Phi}$ and $c = 1$, then Φ is called *orthonormal*. A necessary condition for Φ and $\widetilde{\Phi}$ to be biorthogonal is that their symbols $\mathbf{A}(z)$ and $\widetilde{\mathbf{A}}(z)$ satisfy

$$\sum_{\tilde{\rho} \in \widetilde{R}} \mathbf{A}(z_{M^{-\top}\tilde{\rho}}) \overline{\widetilde{\mathbf{A}}(z_{M^{-\top}\tilde{\rho}})}^\top = m^2 \mathbf{I}_m. \tag{3.18}$$

It was shown in [18] that under mild assumptions this condition is also sufficient as follows.

Theorem 3.2.1. *Let Φ and $\widetilde{\Phi}$ be r–scaling vectors with finitely supported masks $A, \widetilde{A} \in \ell_0(\mathbb{Z}^d)^{r \times r}$. Φ and $\widetilde{\Phi}$ are biorthogonal if and only if the following statements hold:*

(i) $\mathbf{A}(z)$ and $\widetilde{\mathbf{A}}(z)$ satisfy (3.18),

(ii) $1 \in \operatorname{spec}(\mathbf{A}_i(1)^\top)$ and $1 \in \operatorname{spec}(\widetilde{\mathbf{A}}_i(1)^\top)$ for all $0 \leq i < r$,

(iii) $\mathbf{A}(1)$ and $\widetilde{\mathbf{A}}(1)$ have a simple eigenvalue m and all other eigenvalues are smaller than m in modulus, and

(iv) both $\mathbf{T_A}$ and $\mathbf{T_{\widetilde{A}}}$ have a simple eigenvalue 1 and all other eigenvalues are smaller than 1 in modulus.

As mentioned in Section 3.1.5, the spectra of $\mathbf{T_A}$ and $\mathbf{T_{\widetilde{A}}}$ coincide with the spectra of the (finite) matrices $T_{\mathfrak{A},K}$ and $T_{\widetilde{\mathfrak{A}},\widetilde{K}}$, respectively. Therefore, given the masks A and $\widetilde{A}$, these conditions can easily be checked. For the orthonormal case, i.e., $\widetilde{\Phi} = \Phi$, this theorem has already been proven in [75].

Now, let Φ and $\widetilde{\Phi}$ satisfy the assumptions of Theorem 3.18, i.e., $\Phi, \widetilde{\Phi} \in L_2(\mathbb{R}^d)^r$ and $A, \widetilde{A} \in \ell_0(\mathbb{Z}^d)^{r \times r}$. Then the biorthogonality condition (3.17) implies that Φ and $\widetilde{\Phi}$ are ℓ_2–stable, cf. [19, 22]. Moreover, Theorem 3.1.1 yields that both, Φ as well as $\widetilde{\Phi}$, are compactly supported. Consequently, each scaling vector generates an MRA. If there exist some multiwavelets associated to these MRA, then, as we have seen in Section 3.1.3, the task of finding these multiwavelets can be reduced to finding the corresponding symbols. Moreover, it is commonly known that finding the symbols of biorthogonal multiwavelets leads to the following *matrix extension problem*. Let $\mathbf{B}^{(n)}(z)$ denote the symbol of $\Psi^{(n)}$ for $1 \le n < m$. The *modulation matrix* of Φ is defined by

$$
\mathcal{P}_m(z) := \frac{1}{m}
\begin{pmatrix}
A\big(z_{M^{-\top}\tilde{\rho}_0}\big) & \cdots & A\big(z_{M^{-\top}\tilde{\rho}_{m-1}}\big) \\
\mathbf{B}^{(1)}\big(z_{M^{-\top}\tilde{\rho}_0}\big) & \cdots & \mathbf{B}^{(1)}\big(z_{M^{-\top}\tilde{\rho}_{m-1}}\big) \\
\vdots & & \vdots \\
\mathbf{B}^{(m-1)}\big(z_{M^{-\top}\tilde{\rho}_0}\big) & \cdots & \mathbf{B}^{(m-1)}\big(z_{M^{-\top}\tilde{\rho}_{m-1}}\big)
\end{pmatrix}.
\tag{3.19}
$$

For dual multiwavelets $\widetilde{\Psi}^{(n)}$ with symbols $\widetilde{\mathbf{B}}^{(n)}(z)$, define $\widetilde{\mathcal{P}}_m(z)$ analogously. The following theorem has been shown in [22, 104] for $r = 1$. The generalization to our setting is straightforward, see also [12, 34, 74].

Theorem 3.2.2. *Let Φ and $\widetilde{\Phi}$ be a pair of compactly supported biorthogonal scaling vectors with masks $A, \widetilde{A} \in \ell_0(\mathbb{Z}^d)^{r \times r}$, and let $\Psi^{(n)}$ and $\widetilde{\Psi}^{(n)}$, $1 \le n < m$, be defined by (3.12) with masks $B^{(n)}, \widetilde{B}^{(n)} \in \ell_0(\mathbb{Z}^d)^{r \times r}$, respectively. Then $\Psi^{(n)}$ and $\widetilde{\Psi}^{(n)}$ give rise to biorthogonal multiwavelet bases if and only if*

$$
\mathcal{P}_m(z)\overline{\widetilde{\mathcal{P}}_m(z)}^{\top} = \mathbf{I}_{mr}
\tag{3.20}
$$

for all $z \in \mathbb{T}^d$.

Hence, the construction of biorthogonal multiwavelet bases can be decomposed into two steps. First, one has to construct a biorthogonal pair of scaling vectors, and then one has to solve the above extension problem to obtain the corresponding biorthogonal multiwavelets. In the orthonormal case, the modulation matrix $\mathcal{P}_m(z)$ has to be unitary for almost all $z \in \mathbb{T}^d$ to provide an orthonormal multi-wavelet basis. In general, Theorem 3.2.2 can be obtained for weaker assumptions as well. However, in our case the assumptions do perfectly fit.

Unfortunately, we observe that the interpolation property (3.6) and strict orthonormality are incompatible, since the constant c in (3.17) is determined by the length of Φ and $m = |\det(M)|$.

Theorem 3.2.3. *Let $\Phi = (\phi_0, \ldots, \phi_{m-1})^{\top}$ be a compactly supported interpolating m-scaling vector with finite mask that satisfies (3.17) with $\widetilde{\Phi} = \Phi$. Then we have*

$$
\|\phi_i\|_{L_2}^2 = \int_{\mathbb{R}^d} \phi_i(x)\,\mathrm{d}x = \frac{1}{m}
\tag{3.21}
$$

for $i \in \{0, \ldots, m-1\}$.

Proof. For an arbitrary index set $\Lambda \subset \mathbb{Z}^d$ let us define the function

$$\Xi_\Lambda(x) := \sum_{\beta \in \Lambda} \sum_{i=0}^{m-1} \phi_i(x - \beta),$$

and we denote $\Xi := \Xi_{\mathbb{Z}^d}$. Since Φ is compactly supported, there exists a finite index set $\tilde{\Lambda}$ such that the identity $\Xi(x) = \Xi_{\tilde{\Lambda}}(x)$ holds for all $x \in \mathrm{supp}(\Phi)$. Due to (3.17) we have $\|\phi_i\|_{L_2} = \sqrt{c}$ for $i = 0, \ldots, m-1$, thus $c^{-1/2}\Phi$ is an orthonormal scaling vector and we obtain the representation

$$\Xi_{\tilde{\Lambda}}(x) = \sum_{\beta \in \tilde{\Lambda}} \sum_{i=0}^{m-1} \langle \Xi_{\tilde{\Lambda}}, c^{-1/2}\phi_i(x - \beta) \rangle c^{-1/2}\phi_i(x - \beta). \tag{3.22}$$

Furthermore, since Φ is stable, Proposition 3.1.3 implies that $S(\Phi)$ contains the constant functions. Hence, the sampling property (3.8) yields $\Xi \equiv 1$. Therefore, combining (3.8) and (3.22) yields

$$1 = \frac{1}{\sqrt{c}} \langle \Xi_{\tilde{\Lambda}}, \frac{1}{\sqrt{c}}\phi_i \rangle = \frac{1}{\|\phi_i\|_{L_2}^2} \langle 1, \phi_i \rangle.$$

Thus, we have proven the first identity in (3.21).

By definition, Ξ is periodic, and since Φ is compactly supported, we can expand Ξ into its Fourier series. A standard result of Fourier analysis shows that the βth Fourier coefficient $\hat{c}_\Xi(\beta)$ of Ξ satisfies

$$(2\pi)^{-d/2} \hat{c}_\Xi(\beta) = \sum_{i=0}^{m-1} \widehat{\phi_i}(2\pi\beta),$$

see Theorem 5.1 in [107] for details. On the other hand, $\Xi \equiv 1$ implies $\hat{c}_\Xi(\beta) = \delta_{0,\beta}$ and we obtain

$$1 = \sum_{i=0}^{m-1} \int_{\mathbb{R}^d} \phi_i(x)\,\mathrm{d}x.$$

Due to ((3.17)) we have $\|\phi_0\|_{L_2} = \ldots = \|\phi_{m-1}\|_{L_2}$, and with the first identity in (3.21) the proof is complete. $\qquad\square$

Remark 3.2.4. *Although it is impossible for a scaling vector Φ to be interpolating and strictly orthonormal simultaneously, we can switch between these properties via multiplying Φ by $\sqrt{m}$ and $1/\sqrt{m}$ respectively. Thus, in the following, a scaling vector satisfying (3.6) and (3.17) is called an* orthonormal interpolating *scaling vector.*

3.2.2   Approximation Order and Vanishing Moments

For a compactly supported function vector $\Phi \in L_2(\mathbb{R}^d)^r$ and $h > 0$, let

$$S_h(\Phi) := \left\{ f\left(\frac{\cdot}{h}\right) \mid f \in S(\Phi) \cap L_2(\mathbb{R}^d) \right\}$$

denote the space of all h–dilates of $S(\Phi) \cap L_2(\mathbb{R}^d)$. Φ (or $S(\Phi)$) is said to provide *approximation order* $k > 0$ if the Jackson–type inequality

$$\inf_{g \in S_h(\Phi)} \|f - g\|_{L_2} = \mathcal{O}(h^k), \quad \text{as } h \to 0,$$

holds for all f contained in the Sobolev space $H^k(\mathbb{R}^d)$. For an arbitrary $s > 0$, $H^s(\mathbb{R}^d)$ is defined by

$$H^s(\mathbb{R}^d) := \left\{ f \in L_2(\mathbb{R}^d) \mid \int_{\mathbb{R}^d} |\widehat{f}(\xi)|^2 (1 + \|\xi\|_2^s)^2 \, d\xi < \infty \right\}.$$

As in the scalar case, the approximation properties of a scaling vector are closely related to its ability to reproduce polynomials.

A function vector $\Phi : \mathbb{R}^d \longrightarrow \mathbb{C}^r$ with compact support is said to provide *accuracy order* $k + 1$, if $\pi_k^d \subset S(\Phi)$, where π_k^d denotes the space of all polynomials of total degree less or equal than k in $\mathbb{R}^d$. It was shown by Jia, see [69], that if a compactly supported scaling vector Φ has linear independent integer translates or is at least stable, then the order of accuracy is equivalent to the approximation order provided by Φ.

A mask $A \in \ell_0(\mathbb{Z}^d)^{r \times r}$ of an r–scaling vector with respect to a scaling matrix M satisfies the *sum rules of order* k if there exists a set of vectors $\{y_\mu \in \mathbb{R}^r \mid \mu \in \mathbb{Z}_+^d, |\mu| < k\}$ with $y_0 \neq 0$ such that

$$\sum_{0 \leq \nu \leq \mu} (-1)^{|\nu|} \left(\sum_{\beta \in \mathbb{Z}^d} \frac{(M^{-1}\rho + \beta)^\nu}{\nu!} A_{\rho + M\beta}^\top \right) y_{\mu - \nu} = \sum_{|\nu| = |\mu|} w(\mu, \nu) y_\nu \qquad (3.23)$$

holds for all $\mu \in \mathbb{Z}_+^d$ with $|\mu| < k$ and all $\rho \in R$. Here we use standard multi-index notation, i.e., $|\mu| = \mu_1 + \ldots + \mu_d$ and $\mu! = \mu_1! \cdots \mu_d!$. Furthermore, we say $\nu \leq \mu$ if $\nu_i \leq \mu_i$ for all $0 \leq i < d$, and $\nu < \mu$ if $\nu_i < \mu_i$ for at least one $0 \leq i < d$. The numbers $w(\mu, \nu)$ are uniquely determined by

$$\frac{(M^{-1}x)^\mu}{\mu!} = \sum_{|\nu| = |\mu|} w(\mu, \nu) \frac{x^\nu}{\nu!}, \quad \text{for all } x \in \mathbb{R}^d. \qquad (3.24)$$

Remark 3.2.5. *The numbers* $w(\mu, \nu)$ *can also be obtained by the equation*

$$w(\mu, \nu) = D^\nu \frac{(M^{-1}x)^\mu}{\mu!}\bigg|_{x=0}$$

where $D^\nu := \frac{\partial^{|\nu|}}{\partial x_1^{\nu_1} \cdots \partial x_d^{\nu_d}}$.

Remark 3.2.6. *The reader should observe that condition* (ii) *in Theorem 3.2.1 is equivalent to the masks* A *and* $\widetilde{A}$ *satisfying the sum rules of order 1.*

It was proven in [10, 75] that if the mask of a compactly supported scaling vector Φ satisfies the sum rules of order k then Φ provides accuracy of order k. For the univariate case, the sum rules and their correlation with the approximation order and accuracy order of a scaling vector were studied in [98] and [65]. If the mask of a compactly supported scaling vector Φ satisfies the sum rules of order k and, in addition, Φ is stable, then Φ *reproduces polynomials* of total degree less than k, i.e., for all $\mu \in \mathbb{Z}_+^d$ with $|\mu| < k$ we have

$$\frac{x^\mu}{\mu!} = \sum_{0 \le \nu \le \mu} \sum_{\beta \in \mathbb{Z}^d} \frac{\beta^\nu}{\nu!} y_{\mu-\nu}^\top \frac{1}{[\![\Phi]\!]} \Phi(x - \beta) \tag{3.25}$$

with y_μ as above and $[\![\Phi]\!] := \|\widehat{\Phi}(0)\|_2$, see [56, 70] for details.

The reproduction of polynomials by a scaling vector has some impact on the properties of the corresponding multiwavelets as well. Similar to the univariate scalar setting, cf. Section 2.2, a function vector $\Psi \in L_2(\mathbb{R}^d)^r$ is said to have *vanishing moments of order* k, if

$$\langle x^\mu, \Psi(x) \rangle := \left(\langle x^\mu, \psi_0(x) \rangle, \ldots, \langle x^\mu, \psi_{r-1}(x) \rangle \right)^\top = 0$$

for all $\mu \in \mathbb{Z}_+^d$ with $|\mu| < k$. Let $\Psi^{(n)}$, $\widetilde{\Psi}^{(n)}$, $1 \le n < m$, be biorthogonal multiwavelets, and let Φ and $\widetilde{\Phi}$ be the corresponding biorthogonal scaling vectors. As an immediate consequence of the biorthogonality relation (3.14) one obtains

$$\langle \Phi, \widetilde{\Psi}^{(n)}(\cdot - \beta) \rangle = 0 = \langle \widetilde{\Phi}, \Psi^{(n)}(\cdot - \beta) \rangle$$

for all $\beta \in \mathbb{Z}^d$ and $1 \le n < m$. Therefore, Equation (3.25) implies that the the sum rule order of the mask of the dual scaling vector provides a lower bound for the vanishing moment order of the primal multiwavelets. The same relation holds for the dual multiwavelets and the primal scaling vector. On the other hand, as we have already seen in Section 2.2, vanishing moments are the key property for the success of wavelet related algorithms in many applications. For example in signal/image processing, a high order of vanishing moments leads to very small coefficients in the wavelet expansion wherever the signal/image is smooth. Thus, the signal or image can effectively be compressed, see Chapter 7 for a more detailed discussion of this topic.

3.2.3   Regularity

In many fields of application, e.g., image analysis or geometric modeling, the regularity or smoothness of a scaling vector has a strong impact on the performance of the corresponding multiwavelet or subdivision algorithm, respectively. Therefore, this topic has been extensively studied in recent years, see [15, 16, 26, 59, 71, 73, 93] and the references therein. In the following, we focus on the approach derived in [71] which extends the results in [73] to the multivariate setting.

A commonly used measure for the smoothness of a scaling vector $\Phi \in L_2(\mathbb{R}^d)^r$ is the *critical Sobolev exponent*

$$\mathfrak{s}(\Phi) := \sup \left\{ s \,\middle|\, \phi_i \in H^s(\mathbb{R}^d) \text{ for all } i = 1, \dots, r \right\}.$$

For an *isotropic* scaling matrix M, i.e., M is similar to a diagonal matrix given by the diagonal $(\sigma_0, \dots, \sigma_{d-1})$ with $|\sigma_0| = \dots = |\sigma_{d-1}|$, the critical Sobolev exponent $\mathfrak{s}(\Phi)$ is closely related to the spectrum of the transition operator $T_{\mathfrak{A}}$. The following theorem has been proven in [71]. We use the notation of Section 3.1.5.

Theorem 3.2.7. *Let $\Phi \in L_2(\mathbb{R}^d)^r$ be a compactly supported solution of* (3.1) *with mask $A \in \ell_0(\mathbb{Z}^d)^{r \times r}$ and an isotropic scaling matrix M with eigenvalues $\sigma_0, \dots, \sigma_{d-1}$. Define $\sigma := (\sigma_0, \dots, \sigma_{d-1})^\top$ and $\Lambda := \operatorname{spec}\left(\frac{1}{m}\mathbf{A}(1)\right) \setminus \{1\}$. Suppose Φ provides accuracy order k, then with*

$$E_k := \left\{ \lambda \overline{\sigma^{-\mu}}, \overline{\lambda} \sigma^{-\mu} \,\middle|\, \lambda \in \Lambda, |\mu| < k \right\} \cup \left\{ \sigma^{-\mu} \,\middle|\, |\mu| < 2k \right\}$$

we have

$$\mathfrak{s}(\Phi) \geq -\log_{\operatorname{spr}(M)} \sqrt{\operatorname{spr} k}, \tag{3.26}$$

where

$$\operatorname{spr}_k := \max \left\{ |\eta| \,\middle|\, \eta \in \operatorname{spec}(T_{\mathfrak{A},K}) \setminus E_k \right\}.$$

If, in addition, Φ is stable then in (3.26) *equality holds true.*

This theorem shows that the regularity of a scaling vector is closely related to spectral properties of a large matrix. Hence, conditions for high regularity are hard to incorporate into a construction process. On the other hand, Theorem 3.2.7 provides us with a handy method for computing the smoothness order of a scaling vector, given its mask. Therefore, in this work, we will use regularity estimates obtained by an implementation of the above method as a measure of quality for the outcome of our construction.

Remark 3.2.8. *Throughout this work, we focus on multiwavelets with finitely supported masks. Hence, Equation* (3.12) *implies that these multiwavelets possess the same regularity properties as the underlying scaling vectors.*

Chapter 4

Recipe I: The Univariate Case

Before we study interpolating scaling vectors in full generality, it is worth while to investigate the univariate case in detail. Though it is self-evident that the multivariate case already contains the univariate case, from the constructional point of view the univariate setting is somewhat more accessible. This is due to the fact that some univariate mathematical structures do not carry over to the multivariate setting, e.g., univariate polynomials can be factorized, multivariate polynomials in general can not. Hence, the univariate setting provides a wide variety of mathematical tools which enables us to develop a systematic approach to the construction of interpolating scaling vectors. On the other hand, we will see in the sequel that the basic construction principle does not rely on the chosen dimension. Therefore, although we have to adapt some constructional details, the univariate approach acts as a template for the multivariate construction methods in the following chapters.

In this chapter, we focus on the classical case, i.e., we construct orthonormal interpolating 2–scaling vectors for dyadic scaling ($r = M = 2$). Since the number of multiwavelets is determined by the scaling parameter, cf. Section 3.1.3, this approach enables us to construct orthonormal bases of $L_2(\mathbb{R})$ generated by one single mother multiwavelet consisting of two functions only. Furthermore, the choice $M = 2$ helps to maintain the template characteristic of our approach by keeping the constructional complexity at a bearable level. The following results have already been published in [79].

4.1 Main Ingredients

We have seen in Section 3.1.1 that under certain mild conditions a scaling vector is completely determined by its mask or symbol, respectively, up to a constant. Furthermore, for almost all the properties of scaling vectors introduced in Section

3.2 there exist at least necessary conditions in terms of the symbol. Thus, it suggests itself to start our construction process by collecting and simplifying these conditions to design some suitable symbols.

Since we focus on dyadic scaling, the interpolation property (3.6) implies $r = m = 2$. Hence, for the set of representatives of $\mathbb{Z}/2\mathbb{Z}$ we can choose $R = \tilde{R} = \{0, 1\}$. In addition, we only consider masks of finite support, i.e., $A \in \ell_0(\mathbb{Z}^d)^{2\times 2}$. Thus, Theorem 3.1.1 ensures compact support of the corresponding scaling vectors as long as the eigenvalue condition $\operatorname{spec}(\mathbf{A}(1)) = \{2, \lambda \,|\, |\lambda| < 2\}$ is satisfied.

4.1.1   Orthonormality

Theorem 3.2.1 tells us that the orthonormality of a scaling vector with finite mask can be completely characterized in terms of the symbol. In particular, the symbol $\mathbf{A}(z)$ of an orthonormal scaling vector necessarily has to stem from a *conjugate quadrature filter*, i.e., $\mathbf{A}(z)$ has to satisfy

$$\mathbf{A}(z)\overline{\mathbf{A}(z)}^{\mathsf{T}} + \mathbf{A}(-z)\overline{\mathbf{A}(-z)}^{\mathsf{T}} = 4\,\mathbf{I}_2, \quad z \in \mathbb{T}. \tag{4.1}$$

Applying the interpolation condition (3.11), we obtain the following simplification.

Theorem 4.1.1. *Let Φ be an interpolating 2-scaling vector with mask $A \in \ell_0(\mathbb{Z})^{2\times 2}$. With the notation of Equation (3.11), the symbol $\mathbf{A}(z)$ satisfies (4.1) if and only if the symbol entries $a^{(0)}(z)$ and $a^{(1)}(z)$ satisfy*

$$2 = |a^{(0)}(z)|^2 + |a^{(0)}(-z)|^2 \tag{4.2}$$

and

$$a^{(1)}(z) = \pm z^{2\kappa+1}\overline{a^{(0)}(-z)}, \quad \kappa \in \mathbb{Z}, \tag{4.3}$$

with

$$a^{(0)}(z) = \sum_{\beta=\gamma_0}^{\Gamma_0} a_\beta z^\beta, \quad \gamma_0, \Gamma_0 \in \mathbb{Z}, \quad \Gamma_0 - \gamma_0 \text{ odd}.$$

Proof. Let the symbol $\mathbf{A}(z)$ satisfy (4.1). Then the interpolation condition (3.11) leads to

$$\begin{aligned}
2 &= |a^{(0)}(z)|^2 + |a^{(0)}(-z)|^2 & (4.4) \\
2 &= |a^{(1)}(z)|^2 + |a^{(1)}(-z)|^2 & (4.5) \\
0 &= a^{(0)}(z)\overline{a^{(1)}(z)} + a^{(0)}(-z)\overline{a^{(1)}(-z)} & (4.6)
\end{aligned}$$

for all $z \in \mathbb{T}$. Since the mask A has finite support, the symbol entries $a_i(z)$, $i \in \{0, 1\}$, are Laurent polynomials, i.e., there exist $\gamma_i, \Gamma_i \in \mathbb{Z}$ with $\gamma_i \leq \Gamma_i$ such

that
$$a_i(z) = \sum_{\beta=\gamma_i}^{\Gamma_i} a_\beta^{(i)} z^\beta$$

with $a_{\Gamma_i}^{(i)} \neq 0$ and $a_{\gamma_i}^{(i)} \neq 0$. Therefore, there exist polynomials $\check{a}_i(z)$, $\bar{\check{a}}_i(z)$ with

$$a_i(z) = z^{\gamma_i}\check{a}_i(z), \quad a_i(z^{-1}) = z^{-\Gamma_i}\bar{\check{a}}_i(z), \quad i = 0, 1.$$

The coefficients of $a_i(z)$ are real and $z \in \mathbb{T}$, therefore $\overline{a_i(z)} = a_i(z^{-1})$. Thus (4.4) is equivalent to

$$\check{a}^{(0)}(z)\bar{\check{a}}_0(z) + (-1)^{\gamma_0-\Gamma_0}\check{a}^{(0)}(-z)\bar{\check{a}}_0(-z) = 2z^{\Gamma_0-\gamma_0}.$$

Consisting of polynomials only, this equation holds for all $z \in \mathbb{C}$. Since the right hand side is a monomial and $\check{a}^{(0)}(0) = a_{\gamma_0}^{(0)} \neq 0$, we obtain for the greatest common divisor of $\check{a}^{(0)}(z)$ and $\check{a}^{(0)}(-z)$

$$\gcd(\check{a}^{(0)}(z), \check{a}^{(0)}(-z)) = 1.$$

Furthermore, $\Gamma_0 - \gamma_0$ is odd, because $\bar{\check{a}}^{(0)}(0) = a_{\Gamma_0}^{(0)} \neq 0$ as well.

With the above notation Equation (4.6) is equivalent to

$$\check{a}^{(0)}(z)\bar{\check{a}}_1(z) = -(-1)^{\gamma_0-\Gamma_1}\check{a}^{(0)}(-z)\bar{\check{a}}_1(-z),$$

which also holds for all $z \in \mathbb{C}$. Comparing the linear factors of the polynomials on both sides, we obtain that $\bar{\check{a}}_1(z)$ has to contain the linear factors which are contained in $\check{a}^{(0)}(-z)$ and which are not contained in $\check{a}^{(0)}(z)$. Therefore, $\bar{\check{a}}_1(z)$ has to be of the form

$$\bar{\check{a}}_1(z) = \frac{\check{a}^{(0)}(-z)}{\gcd(\check{a}^{(0)}(z), \check{a}^{(0)}(-z))}p(z) = \check{a}^{(0)}(-z)p(z) \qquad (4.7)$$

with a polynomial $p(z)$. Applying this to (4.6) yields

$$p(z) = -(-1)^{\gamma_0-\Gamma_1}p(-z)$$

and by Equation (4.5) there exists an $\alpha \in \mathbb{Z}$ such that

$$p(z) = \pm z^{\Gamma_1-\gamma_0+2\alpha+1}.$$

Yet, because of $\bar{\check{a}}_1(0) = a_{\Gamma_1}^{(1)} \neq 0$, Equation (4.7) yields $p(z) \equiv \pm 1$ and $\Gamma_1 - \gamma_0$ is odd. Therefore $a^{(1)}(z)$ has to be of the form

$$a^{(1)}(z) = \pm z^{\Gamma_1-\gamma_0}\overline{a^{(0)}(-z)} =: \pm z^{2\kappa+1}\overline{a^{(0)}(-z)}.$$

On the other hand, if the symbol entry $a^{(0)}(z)$ satisfies (4.2) and $a^{(1)}(z)$ is defined via (4.3) then a short computation shows that (4.1) is satisfied as well. □

Remark 4.1.2. *If $a^{(0)}(z)$ is a (Laurent-) monomial, then due to the above equations the same holds for $a^{(1)}(z)$ with*

$$a^{(0)}(z) = z^\alpha, \quad a^{(1)}(z) = z^{\alpha+2\kappa+1}, \quad \alpha, \kappa \in \mathbb{Z}.$$

By Theorem 1.2 in [47] the corresponding scaling vector is of Haar–type, i. e., it is the characteristic function of a self-affine multi-tile. Therefore, this scaling vector can not even be continuous.

The next step is to find a more applicable version of condition (4.2). First of all, similar to the construction of the orthonormal Daubechies wavelets in [36] we represent $|a^{(0)}(z)|^2$, $z = e^{-i\omega}$, as a polynomial in $\sin^2 \frac{\omega}{2}$

$$|a^{(0)}(e^{-i\omega})|^2 =: P\left(\sin^2 \frac{\omega}{2}\right).$$

Then Equation (4.2) transforms to

$$2 = P\left(\sin^2 \frac{\omega}{2}\right) + P\left(1 - \sin^2 \frac{\omega}{2}\right) =: \sum_{\beta=0}^{\Gamma_0-\gamma_0} p_\beta \left(\sin^2 \frac{\omega}{2}\right)^\beta$$

and we obtain $p_\beta = 2\delta_{0,\beta}$, $\beta = 0 \ldots \Gamma_0 - \gamma_0$. A simple computation yields the following corollary.

Corollary 4.1.3. *Let $a^{(0)}(z)$ be a Laurent polynomial defined on the unit circle via*

$$a^{(0)}(z) := \sum_{\beta=\gamma_0}^{\Gamma_0} a_\beta z^\beta, \quad z \in \mathbb{T}.$$

Then, with $K := \Gamma_0 - \gamma_0$, Equation (4.2) is equivalent to

$$(a_{\gamma_0}, \ldots, a_{\Gamma_0}) \mathbf{M}_\alpha \begin{pmatrix} a_{\gamma_0} \\ \vdots \\ a_{\Gamma_0} \end{pmatrix} = \delta_{0,\alpha} \quad \text{for} \quad \alpha = 0, \ldots, \left\lfloor \frac{K}{2} \right\rfloor, \tag{4.8}$$

with $(K+1) \times (K+1)$ matrices

$$\mathbf{M}_\alpha := \sum_{\beta=\alpha}^{\lfloor K/2 \rfloor} c_{\beta,\alpha} \begin{pmatrix} 0 & \mathbf{I}_{K-2\beta+1} \\ 0 & 0 \end{pmatrix}$$

and

$$c_{\beta,\alpha} := 2^{\delta_{0,\alpha}(1-\delta_{\beta,0})} \sum_{\mu=0}^{\beta-\alpha} \sum_{\nu=\mu}^{\beta} \binom{2\beta}{2\nu} \binom{\nu}{\nu-\mu}.$$

4.1.2 Sum Rules

In this section, we investigate conditions for approximation order and accuracy of a scaling vector. As we have seen in Section 3.2.2, these characteristics are determined by the sum rule order of the corresponding mask. On the other hand, in the scalar case ($r = 1$) accuracy order is also connected with a specific factorization of the symbol which can be used for the smoothness analysis of the scaling function. In [26, 98] it has been shown that this holds for the case $r > 1$ as well. Let D denote the differential operator with respect to ω in terms of $z = e^{-i\omega}$, i.e., $DA(e^{-i\omega}) := \left(\frac{d}{d\omega}A(e^{-i\cdot})\right)(\omega)$. The following theorem, stated in [98], provides us with a univariate version of the sum rules and the corresponding factorization.

Theorem 4.1.4. *Let Φ be a compactly supported r-scaling vector with linearly independent integer translates. Then Φ provides approximation order k if and only if the symbol $\mathbf{A}(z)$ of Φ satisfies the following conditions:*
The elements of $\mathbf{A}(z)$ are Laurent polynomials, and there are vectors $\widehat{y}_\mu \in \mathbb{R}^r, \widehat{y}_0 \neq 0, \mu = 0, \ldots, k - 1$, such that for $\mu = 0, \ldots, k - 1$

$$
\begin{aligned}
\sum_{\nu=0}^{\mu} \binom{\mu}{\nu} \widehat{y}_{\mu-\nu}^{\top} (2i)^{-\nu} \left(D^\nu \mathbf{A}\right)(1) &= 2^{1-\mu}\widehat{y}_\mu^{\top}, \\
\sum_{\nu=0}^{\mu} \binom{\mu}{\nu} \widehat{y}_{\mu-\nu}^{\top} (2i)^{-\nu} \left(D^\nu \mathbf{A}\right)(-1) &= 0
\end{aligned}
\tag{4.9}
$$

holds. Furthermore there exist matrices of Laurent polynomials $\mathbf{C}_\mu(z)$ for $\mu = 0, \ldots, k - 1$, such that $\mathbf{A}(z)$ factorizes like

$$
\mathbf{A}(z) = \frac{1}{2^{m-1}} \mathbf{C}_0(z^2) \cdots \mathbf{C}_{k-1}(z^2) \mathbf{A}_k(z) \mathbf{C}_{k-1}^{-1}(z) \cdots \mathbf{C}_0^{-1}(z),
\tag{4.10}
$$

where $\mathbf{A}_k(z)$ is a suitable matrix with Laurent polynomials as entries.

Remark 4.1.5. *A straightforward computation shows that the Equations (4.9) are equivalent to the sum rules (3.23) with $y_\mu = \frac{1}{\mu!}\widehat{y}_\mu$.*

In the scalar case, a factorization of the symbol can be profitably used for the construction of scaling functions even in the multivariate case, cf. [31]. For the vector setting this subject is somewhat more involved. Since within the factorization the entries of all participating matrices are shuffled, the properties of a single entry of the symbol can hardly be controlled. Thus, it is difficult to incorporate the conditions for interpolation (3.11) and for orthonormality provided by Theorem 4.1.1 into a construction method based on factorization. Furthermore, it is not known yet if such a factorization makes sense or at least exists in the multivariate case. On the other hand, the Equations (4.9), i.e., an equivalent of the sum rules in terms of the symbol, are well suited for construction purposes. In the following theorem we show that the specific form (3.11) of the symbol determines the vectors $\widehat{y}_\mu$ up to a constant, and therefore leads to remarkable simplifications concerning the Equations (4.9).

Theorem 4.1.6. *Let Φ be a compactly supported interpolating 2-scaling vector with symbol $\mathbf{A}(z)$ as in (3.11). For $n \geq 0$ and $i \in \{0,1\}$ define the functions*

$$
d_i^{(n)}(z) := \begin{cases} a_i(z), & \text{if } n = 0, \\[2mm] \displaystyle\sum_{\beta \neq 0} \beta^n a_\beta^{(i,1)} z^\beta, & \text{else.} \end{cases}
$$

Then Φ provides approximation order k if and only if $a_0(z)$ and $a_1(z)$ are Laurent polynomials and for $\mu = 0, \ldots, k-1$ the equations

$$
\begin{aligned}
2^{1-\mu} &= (-1)^n d_0^{(\mu)}(1) + \sum_{\nu=0}^{\mu} \binom{\mu}{\nu}(-1)^\nu d_1^{(\nu)}(1), \\
0 &= (-1)^\mu d_0^{(\mu)}(-1) + \sum_{\nu=0}^{\mu} \binom{\mu}{\nu}(-1)^\nu d_1^{(\nu)}(-1)
\end{aligned}
$$
(4.11)

hold.

Proof. As stated in Section 3.1.2, the interpolation property implies linearly independent integer translates, therefore the hypotheses of Theorem 4.1.4 are satisfied. In the following, we show the equivalence of the sum rules (4.9) and the Equations (4.11). For $\mu \geq 0$ the derivatives $D^\mu \mathbf{A}(z)$ have the form

$$
D^\mu \mathbf{A}(z) = \begin{pmatrix} 0 & (-i)^\mu d_0^{(\mu)}(z) \\[2mm] (-i)^\mu z & (-i)^\mu d_1^{(\mu)}(z) \end{pmatrix}.
$$

With $\widehat{y}_\mu := (\widehat{y}_\mu^{(1)}, \widehat{y}_\mu^{(2)})^\top$ the first columns of the Equations (4.9) are equivalent to

$$
\begin{aligned}
2^{1-\mu}\widehat{y}_\mu^{(1)} &= \widehat{y}_\mu^{(1)} + \sum_{\nu=0}^{\mu} \binom{\mu}{\nu}(-2)^{-\nu} \widehat{y}_{\mu-\nu}^{(2)}, \\
0 &= \widehat{y}_\mu^{(1)} - \sum_{\nu=0}^{\mu} \binom{\mu}{\nu}(-2)^{-\nu} \widehat{y}_{\mu-\nu}^{(2)}.
\end{aligned}
$$

By induction we obtain

$$
\widehat{y}_\mu = \widehat{y}_0^{(1)} \begin{pmatrix} \delta_{0,\mu} \\ 2^{-\mu} \end{pmatrix}
$$
(4.12)

with some $\widehat{y}_0^{(1)} \neq 0$. The second columns of the Equations (4.9) are equivalent to

$$2^{1-\mu}\widehat{y}_\mu^{(2)} = \sum_{\nu=0}^{\mu} \binom{\mu}{\nu}(2i)^{-\nu}\left((-i)^\nu \widehat{y}_{\mu-\nu}^{(1)}a_0^{(\nu)}(1) + (-i)^\nu \widehat{y}_{\mu-\nu}^{(2)}a_1^{(\nu)}(1)\right),$$

$$0 = \sum_{\nu=0}^{\mu} \binom{\mu}{\nu}(2i)^{-\nu}\left((-i)^\nu \widehat{y}_{\mu-\nu}^{(1)}a_0^{(\nu)}(-1) + (-i)^\nu \widehat{y}_{\mu-\nu}^{(2)}a_1^{(\nu)}(-1)\right).$$

Applying (4.12) we obtain (4.11). □

Remark 4.1.7. *The corresponding factorization matrices* $\mathbf{C}_\mu$ *have the form*

$$\mathbf{C}_\mu(z) = \left(\frac{1}{2\widehat{y}_0^{(1)}}\right)^{\delta_{0,\mu}} \begin{pmatrix} 2 & -2 \\ -2z & 2 \end{pmatrix}, \tag{4.13}$$

as they are completely determined by the vectors $\widehat{y}_\mu$, *see* [98] *for details concerning their construction.*

As stated above, linearly independent integer translates of a scaling vector are implied by the interpolation property. Thus, applying the orthonormality condition (4.3) to the sum rules (4.11) yields:

Corollary 4.1.8. *With the notations and conditions of Theorem 4.1.1 an orthonormal interpolating 2-scaling vector* Φ *provides approximation order* k *if and only if for* $\mu = 0, \ldots, k-1$ *the mask coefficients of the symbol entry* $a_0(z)$ *satisfy*

$$2^{-\mu} = \sum_\beta (-2\beta)^\mu a_{2\beta} - \sum_\beta (2(\beta - \kappa) + 1)^\mu a_{2\beta+1},$$

$$2^{-\mu} = \sum_\beta (-2\beta - 1)^\mu a_{2\beta+1} + \sum_\beta (2(\beta - \kappa))^\mu a_{2\beta}. \tag{4.14}$$

The eigenvalue properties of $\mathbf{A}(z)$ provided by Theorem 3.1.1 are likewise simplified by the specific structure of $\mathbf{A}(z)$. To provide a simple eigenvalue 2 for $z = 1$ the entries of $\mathbf{A}(z)$ have to satisfy

$$a_0(1) + a_1(1) = 2 \tag{4.15}$$

which corresponds to the first equation in (4.11) with $\mu = 0$. The second eigenvalue λ of $\mathbf{A}(1)$ is given by

$$a_0(1) = (\lambda - 1)(\lambda - a_1(1)). \tag{4.16}$$

Note that the sum rules of order one together with the orthonormality condition (4.3) already imply that $\mathbf{A}(1)$ has the eigenvalues 2 and 0.

4.2   Explicit Construction

4.2.1   General Method

Based on the results in the preceding section we suggest the following construction principle for the symbol $\mathbf{A}(z)$ of an orthonormal interpolating 2-scaling vector:

1. Start with a first symbol entry $a_0(z)$ with an even number of coefficients

$$a_0(z) := \sum_{\beta=-n+\alpha}^{n+1+\alpha} a_\beta z^\beta, \qquad \alpha \in \mathbb{Z},$$

   by choosing the length $2n + 2$ of $(a_\beta)_{\beta \in \mathbb{Z}}$. Centering the coefficients around a_0 seems to provide the highest regularity, therefore $\alpha = 0$ is chosen.

2. By Equation (4.3) the second symbol entry $a_1(z)$ has to be of the form

$$a_1(z) := \pm z^{2\kappa+1} a_0\left(-z^{-1}\right).$$

   It turns out that the choice $\kappa = 0$ and a positive sign provide the highest regularity and the shortest support. Now we have $2n$ degrees of freedom.

3. Apply the orthonormality condition (4.8) to the coefficient sequence $(a_\beta)_{\beta \in \mathbb{Z}}$. Then we are left with n degrees of freedom.

4. Finally, apply the sum rules (4.14) up to the highest possible order to the coefficient sequence $(a_\beta)_{\beta \in \mathbb{Z}}$.

The steps 3 and 4 of this method yield a system of quadratic and linear equations in $(a_\beta)_{-n \leq \beta \leq n+1}$. In most cases, i.e., if n is not too large, this system can be solved analytically using a symbolic computation tool like Maple or MuPAD. The symbols being constructed in this way correspond to 2-scaling vectors supported on the interval $[-n, n+1]$. However, since the applied conditions are just necessary, we have to check whether the scaling vectors really possess the desired properties or not. On the other hand, since the sum rules appear pairwise, the system of equations may be underdetermined. Moreover, due to the quadratic conditions involved in our construction method, there might exist several solutions anyway. Therefore, after solving the system of equations, a screening process has to be employed which extracts solutions with most appealing properties. For this purpose, regularity estimation, e.g., by means of the method described in Section 3.2.3, is an appropriate tool. We will discuss this topic in detail in Section 5.2 for the more general multivariate case.

4.2.2 Examples

In the sequel, we present the outcome of our construction for $n = 1, \ldots, 8$. As stated above, the solutions obtained by our construction are not unique. Therefore, we focus on those solutions which possess the highest regularity.

The case $n = 1$:

For the case $n = 1$, our construction leads to a one-parameter set of 2-scaling vectors Φ_α with symbols

$$\mathbf{A}_\alpha(z) = \begin{pmatrix} 1 & \sqrt{-\alpha(\alpha-1)}z^{-1} + \alpha - \sqrt{-\alpha(\alpha-1)}z + (1-\alpha)z^2 \\ z & (1-\alpha)z^{-1} + \sqrt{-\alpha(\alpha-1)} + \alpha z - \sqrt{-\alpha(\alpha-1)}z^2 \end{pmatrix}.$$

Figure 4.1 shows the critical Sobolev exponent $\mathfrak{s}(\Phi_\alpha)$. For $\alpha \approx 0.9486$ we obtain

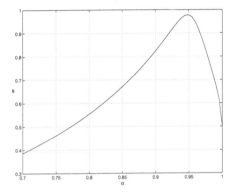

Figure 4.1: Sobolev exponent $\mathfrak{s}$ of Φ_α

an interpolating and orthonormal 2-scaling vector depicted in Figure 4.2(a) which provides approximation order 1 and is supported on $[-1, 2]$. The critical Sobolev exponent of $\Phi_{0.9486}$ is $\mathfrak{s} = 0.9777$. For $\alpha = 1$ we also obtain the Haar generator interpreted as a scaling vector via (3.7) as is shown in Figure 4.2(b).

The case $n > 1$:

As stated above, for each fixed input parameter n the outcome of our construction process is likely to be not unique. In contrast to the case $n = 1$, it turns our that for $n > 1$ we do not obtain a parameter depending family but a discrete set of solutions. In this example we concentrate on the family Φ_n of the most regular

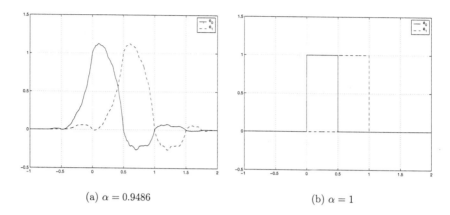

(a) $\alpha = 0.9486$ (b) $\alpha = 1$

Figure 4.2: 2-Scaling vector Φ_α

elements of these sets. All these Φ_n depicted in Figure 4.3 are very similar in shape. The main mass of the scaling vector is concentrated in the interval $[-1, 2]$ and with increasing n there is just some oscillation added outside this interval. The coefficient sequences $(a_\beta)_{\beta \in \mathbb{Z}}$ of the corresponding symbol entries $a_0(z)$ are listed in Appendix A.1. They also reveal this similarity.

n	supp Φ	approximation order k	Sobolev exponent $\mathfrak{s}$
2	[-2,3]	2	1.50
3	[-3,4]	3	1.51
4	[-4,5]	3	1.74
5	[-5,6]	4	1.80
6	[-6,7]	4	2.01
7	[-7,8]	5	1.84
8	[-8,9]	5	2.04

Table 4.1: Sobolev regularity and approximation order of Φ_n

As is shown in Table 4.1, similar to the non-orthonormal case the Sobolev regularity and the provided approximation order increase with the support length of Φ_n. By enlarging the parameter n by one either the critical Sobolev exponent of Φ_n or the provided approximation order is alternately increased. In our examples the approximation order equals $\lfloor \frac{n+1}{2} \rfloor + 1$. Note that the examples for $n = 2$ and

$n = 4$ were also obtained by Selesnick in [109]. For $n = 3$ our construction also yields the corresponding example of Selesnick and, in addition, a scaling vector that provides a higher Sobolev regularity. Therefore the more regular scaling vector is depicted in Figure 4.3.

4.3 Multiwavelets

In the scalar case, the construction of wavelets corresponding to orthonormal scaling functions is by now well-understood due to the pioneering work of Daubechies in [36]. All possible wavelets are related to a canonical wavelet, which is completely determined by the symbol of the scaling function. In the vector case the situation is somewhat more involved.

We have seen in Section 3.2.1 that for a compactly supported orthonormal 2–scaling vector Φ, which consequently generates an MRA, the task of finding a multiwavelet Ψ can be converted into a matrix extension problem. Let $\mathbf{A}(z)$ be the symbol of Φ, then we have to find a symbol $\mathbf{B}(z)$ such that the modulation matrix $\mathcal{P}_m(z)$ defined in (3.20) is unitary for all $z \in \mathbb{T}$. Unlike the scalar case, there may exist several unitary extensions, and thus the construction of multiwavelets is by no means unique.

One possibility to obtain $\mathbf{B}(z)$ is to use the results in [82] which provide us with an effective algorithm for the extension of $\mathcal{P}_m(z)$. Another possibility is to impose some additional conditions on the multiwavelet Ψ or its symbol $\mathbf{B}(z)$. Since our scaling vectors are interpolating, we may choose the multiwavelet Ψ to be interpolating as well, i. e.,

$$\Psi\left(\frac{n}{2}\right) = \begin{pmatrix} \delta_{0,n} \\ \delta_{1,n} \end{pmatrix}.$$

Then our construction leaves us much less freedom in the sense that the symbol $\mathbf{B}(z)$ of Ψ is completely determined by the symbol $\mathbf{A}(z)$ of Φ. The following theorem has already been stated in [109] but without a proof. For the reader's convenience we sketch the proof in our setting.

Corollary 4.3.1. *The symbol $\mathbf{B}(z)$ of an compactly supported interpolating multiwavelet corresponding to a compactly supported orthonormal interpolating 2-scaling vector with symbol $\mathbf{A}(z)$ as in Theorem 4.1.1 has to satisfy*

$$\mathbf{B}(z) = \begin{pmatrix} 1 & -a_0(z) \\ z & -a_1(z) \end{pmatrix}, \quad z \in \mathbb{T}.$$

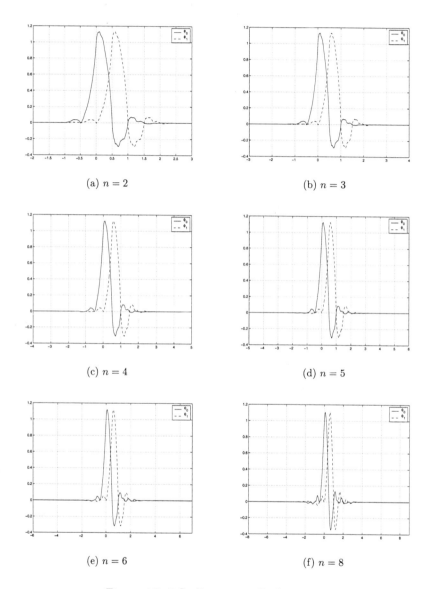

(a) $n = 2$ (b) $n = 3$

(c) $n = 4$ (d) $n = 5$

(e) $n = 6$ (f) $n = 8$

Figure 4.3: 2-Scaling vector Φ_n for $n > 1$

Proof. Due to the interpolation property $\mathbf{B}(z)$ has to be of the form

$$\mathbf{B}(z) = \begin{pmatrix} 1 & b_0(z) \\ z & b_1(z) \end{pmatrix}, \quad z \in \mathbb{T},$$

with Laurent polynomials $b_0(z)$, $b_1(z)$. The unitarity of $\mathcal{P}_m(z)$ leads to

$$\begin{aligned} -2 &= a_i(z)\overline{b_i(z)} + a_i(-z)\overline{b_i(-z)} \\ 0 &= a_i(z)\overline{b_{1-i}(z)} + a_i(-z)\overline{b_{1-i}(-z)} \\ 2 &= |b_i(z)|^2 + |b_i(-z)|^2 \end{aligned}$$

for $i = 0, 1$. Following the lines of the proof of Theorem 4.1.1 we obtain

$$b_0(z) = -a_0(z) \quad \text{and} \quad b_1(z) = -a_1(z).$$

$\square$

Thus, we have two distinct methods to compute interpolating as well as non-interpolating multiwavelets corresponding to our interpolating scaling vectors. In Figure 4.4 both multiwavelets corresponding to $\Phi_{0.9486}$ for $n = 1$ are shown. Note that they possess the same support properties but are distinct in shape.

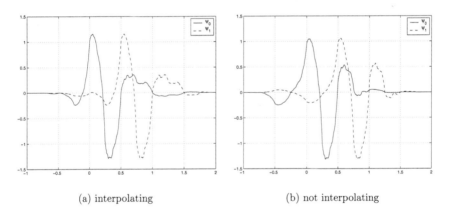

(a) interpolating   (b) not interpolating

Figure 4.4: Multiwavelets corresponding to $\Phi_{0.9486}$

The non-interpolating and interpolating multiwavelets Ψ_n corresponding to the Φ_n for $n > 1$ are depicted in Figure 4.5 and Figure 4.6, respectively. Both multiwavelet families possess rather similar support properties, but the non-interpolating multiwavelets reveal stronger oscillations. The masks of the non-interpolating multiwavelets can be found in Appendix A.1. Again, these masks also reveal the similarity of the multiwavelets shown in Figure 4.5.

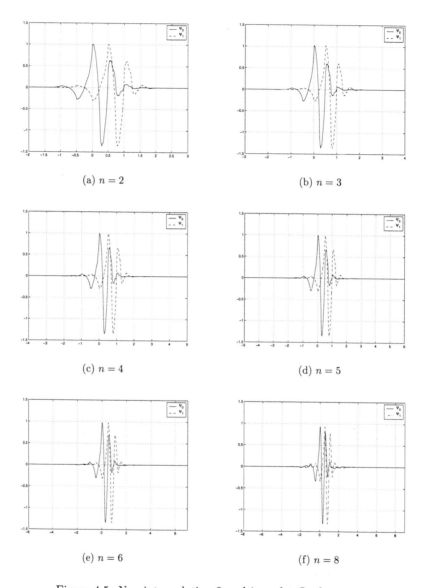

(a) $n = 2$

(b) $n = 3$

(c) $n = 4$

(d) $n = 5$

(e) $n = 6$

(f) $n = 8$

Figure 4.5: Non-interpolating 2-multiwavelet Ψ_n for $n > 1$

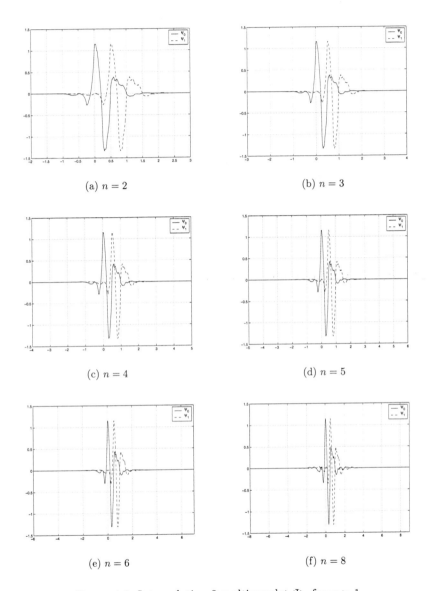

(a) $n = 2$ (b) $n = 3$

(c) $n = 4$ (d) $n = 5$

(e) $n = 6$ (f) $n = 8$

Figure 4.6: Interpolating 2-multiwavelet Ψ_n for $n > 1$

Chapter 5

Recipe II: Multivariate Orthonormal Interpolating Scaling Vectors

In this chapter, we generalize the approach derived in the preceding chapter to the multivariate case. In particular, we focus on scaling matrices with $|\det(M)| = 2$ which enables us to obtain a multivariate analog of Theorem 4.1.1. Similar to the univariate approach, we start by collecting necessary conditions for orthonormality and approximation order of a scaling vector in terms of its mask or symbol, respectively. This enables us to set up an algorithm for the construction of orthonormal interpolating scaling vectors with compact support. Since this algorithm involves solving large nonlinear equation systems, also numerical issues are addressed.

The results within this chapter have already been published in [80].

5.1 Main Ingredients

5.1.1 Orthonormality

As in the univariate case, the symbol $\mathbf{A}(z)$ of an orthonormal scaling vector has to stem from a conjugate quadrature filter, and thus

$$\sum_{\tilde{\rho} \in \tilde{R}} \mathbf{A}\left(z_{M^{-\top}\tilde{\rho}}\right) \overline{\mathbf{A}\left(z_{M^{-\top}\tilde{\rho}}\right)}^{\top} = m^2 \, \mathbf{I}_r, \tag{5.1}$$

cf. Equation (4.1). For the special case of an interpolating 2–scaling vector with compact support we obtain the following simplified conditions.

Theorem 5.1.1. *Let* $\mathbf{A}(z)$ *be the symbol of an interpolating 2–scaling vector with mask* $A \in \ell_0(\mathbb{Z}^d)^{2\times 2}$. $\mathbf{A}(z)$ *satisfies* (5.1) *if and only if the symbol entries* $a^{(0)}(z)$ *and* $a^{(1)}(z)$ *in* (3.11) *satisfy*

$$\left|a^{(0)}(z)\right|^2 + \left|a^{(0)}\left(z_{M^{-\top}\tilde{\rho}}\right)\right|^2 = 2 \tag{5.2}$$

and

$$a^{(1)}(z) = \pm z^\alpha \overline{a^{(0)}(z_{M^{-\top}\tilde{\rho}})} \tag{5.3}$$

for some $\alpha \in [\rho]$ *and with* $\widetilde{R} = \{0, \tilde{\rho}\}$.

In order to prove Theorem 5.1.1 we first have to state the following lemma.

Lemma 5.1.2. *Let* p *be a Laurent polynomial that satisfies*

$$|p(z)| = 1 \quad \text{for all } z \in \mathbb{T}^d. \tag{5.4}$$

Then $p(z) = \pm z^\alpha$ *for some* $\alpha \in \mathbb{Z}^d$.

Proof. Let $(p_\beta)_\beta \in \ell_0(\mathbb{Z}^d)$ denote the coefficient sequence of p, and define the convex polytope $K := \operatorname{conv}\{\beta \in \mathbb{Z}^d \,|\, p_\beta \neq 0\}$ where conv denotes the convex hull. A direct computation shows that (5.4) is equivalent to

$$\sum_{\beta \in K \cap \mathbb{Z}^d} p_\beta \overline{p_{\beta-\gamma}} = \delta_{0,\gamma}, \quad \text{for all } \gamma \in \mathbb{Z}^d. \tag{5.5}$$

Now assume that $p(z) \neq \pm z^\alpha$, then it follows immediately that the set of vertices $E \subset \mathbb{Z}^d$ of K satisfies $|E| > 1$. Therefore, for an arbitrarily chosen $\beta_0 \in E$, the polytope $K' := \operatorname{conv}(E\backslash\{\beta_0\})$ is nonempty, and we have $\beta_0 \notin K'$. Thus, the Hahn–Banach Separation Theorem implies that there exists a hyperplane which separates K' from $\{\beta_0\}$, i.e., there exist $n \in \mathbb{R}^n$ and $\theta \in \mathbb{R}$ such that $n^\top x < \theta < n^\top \beta_0$ for all $x \in K'$. Since each point in K is a convex combination of the vertices in E, we obtain $n^\top x < n^\top \beta_0$ for all $x \in K \backslash \{\beta_0\}$. On the other hand, since K is a (compact) polytope, there exists a $\beta_1 \in E \backslash \{\beta_0\}$ such that $n^\top \beta_1 \leq n^\top x$ for all $x \in K$. Thus, for $x \in K$ we obtain

$$n^\top(x + \beta_0 - \beta_1) = n^\top x - n^\top \beta_1 + n^\top \beta_0 \geq n^\top \beta_0,$$

which implies $K \cap (K + \beta_0 - \beta_1) = \{\beta_0\}$. Therefore, with $\gamma := \beta_1 - \beta_0$, Equation (5.5) yields

$$0 = \sum_\beta p_\beta \overline{p_{\beta+\beta_0-\beta_1}} = p_{\beta_1} \overline{p_{\beta_0}}.$$

As a consequence, we obtain $p_{\beta_0} = 0$ or $p_{\beta_1} = 0$ which contradicts the assumption $p(z) \neq \pm z^\alpha$. $\qquad\square$

Now we can prove Theorem 5.1.1.

Proof of Theorem 5.1.1. With the interpolation condition (3.11), Equation (5.1) is equivalent to the matrix

$$\mathcal{A}(z) := \frac{1}{\sqrt{2}} \begin{pmatrix} a^{(0)}(z) & a^{(0)}(z_{M^{-\top}\tilde{\rho}}) \\ a^{(1)}(z) & a^{(1)}(z_{M^{-\top}\tilde{\rho}}) \end{pmatrix} \tag{5.6}$$

being unitary for all $z \in \mathbb{T}^d$. Therefore, $a^{(0)}(z)$ has to satisfy (5.2) and we have $|\det(\mathcal{A}(z))| = 1$ for all $z \in \mathbb{T}^d$. Since $|\det(\mathcal{A}(z))|$ is a Laurent polynomial, Lemma 5.1.2 implies

$$\det(\mathcal{A}(z)) = \pm z^\alpha$$

for some $\alpha \in \mathbb{Z}^d$. By Cramer's rule we obtain

$$a^{(1)}(z) = \pm z^\alpha \overline{a^{(0)}\left(z_{M^{-\top}\tilde{\rho}}\right)} \tag{5.7}$$

and

$$a^{(1)}\left(z_{M^{-\top}\tilde{\rho}}\right) = \mp z^\alpha \overline{a^{(0)}(z)}. \tag{5.8}$$

Hence, we have (5.3). It remains to be shown that $\alpha \in [\rho]$. Assume $\alpha \in [0]$, then we have $z_{M^{-\top}\tilde{\rho}}^\alpha = z^\alpha$ since there exists a $\beta \in \mathbb{Z}^d$ such that $\alpha = M\beta$ and thus

$$z_{M^{-\top}\tilde{\rho}}^{M\beta} = e^{-i\langle \omega + 2\pi M^{-\top}\tilde{\rho}, M\beta \rangle}$$
$$= e^{-i\langle \omega, M\beta \rangle} e^{-i2\pi\langle \tilde{\rho}, \beta \rangle} = z^{M\beta}.$$

Moreover, Cramer's rule implies $M^{-\top}\tilde{\rho} \in \frac{1}{m}\mathbb{Z}^d$, and consequently

$$\left(z_{M^{-\top}\tilde{\rho}}\right)_{M^{-\top}\tilde{\rho}} = z.$$

Hence, for $\alpha \in [0]$ the Equations (5.7) and (5.8) are incompatible. However, Lemma 3.1.2 yields $z_{M^{-\top}\tilde{\rho}}^\rho = -z^\rho$ which implies $z_{M^{-\top}\tilde{\rho}}^\alpha = -z^\alpha$ for $\alpha \in [\rho]$. As a consequence, for $\alpha \in [\rho]$ the Equations (5.7) and (5.8) are equivalent. On the other hand, if (5.2) and (5.3) are satisfied, then $\mathcal{A}(z)$ is unitary. $\square$

A simple computation yields the following corollary which provides us with the corresponding conditions in terms of the mask.

Corollary 5.1.3. *With the notation $a^{(0)}(z) := \sum_{\beta \in \mathbb{Z}^d} a_\beta z^\beta$ it holds that*

(a) condition (5.2) is equivalent to

$$\sum_{\beta \in \mathbb{Z}^d} a_\beta a_{\beta - M\gamma} = \delta_{0,\gamma} \quad \text{for all } \gamma \in \mathbb{Z}^d \tag{5.9}$$

and

(b) condition (5.3) is equivalent to

$$a^{(1)}(z) = \pm z^\alpha \sum_{\beta \in \mathbb{Z}^d} (-1)^{\mathbb{1}_{[\rho]}(\beta)} a_\beta z^{-\beta}. \qquad (5.10)$$

Thus, the orthonormality of a scaling vector leads to simple quadratic conditions in terms of the mask. On the other hand, from the constructional point of view, orthonormality significantly reduces the number of degrees of freedom by determining the symbol entry $a^{(1)}(z)$ up to the factor $\pm z^\alpha$. This is highly welcome, since the size of the equation system involved in the construction process is considerably reduced.

In the sequel of this section, we give a short proof for the nonexistence of an orthonormal interpolating scaling function with compact support for scaling matrices with determinant ± 2. Although this fact seems to be commonly known within the wavelet community, up to the author's knowledge there is no proof available in the literature. Hence, for the reader's convenience, we sketch the proof in our setting.

Theorem 5.1.4. *Let φ be a compactly supported (a, M)–refinable function with orthonormal integer translates and $|\det(M)| = 2$. If φ satisfies the interpolation condition (2.12), then it is of Haar–type, i.e., it is the characteristic funciton of a self-affine tile and therefore not continuous.*

Proof. First of all, due to the orthogonality of φ, the refinement equation implies $a_\beta = m\langle \varphi, \varphi(M \cdot -\beta) \rangle$ for $\beta \in \mathbb{Z}^d$. Thus, since φ is compactly supported, the symbol $a(z)$ is a Laurent polynomial. From the interpolaton condition (2.12) we immediately obtain that

$$a_{M\beta} = \delta_{0,\beta} \qquad (5.11)$$

for all $\beta \in \mathbb{Z}^d$. Hence, for the first subsymbol $a_0(z)$ of $a(z)$ we obtain $a_0(z) = 1$. Furthermore, the scalar version of the orthonormality condition (5.1) implies that

$$4 = |a(z)|^2 + |a(z_{M^{-\top}\tilde{\rho}})|^2, \qquad (5.12)$$

where $\tilde{\rho}$ denotes the usual nontrivial representative of $\mathbb{Z}^d/M^\top\mathbb{Z}^d$. Using subsymbol notation, we have $a(z) = a_0(z^M) + z^\rho a_1(z^M)$ and $a(z_{M^{-\top}\tilde{\rho}}) = a_0(z^M) - z^\rho a_1(z^M)$, cf. Section 3.1.1. Therefore, Equation (5.12) is equivalent to

$$2 = |a_0(z^M)|^2 + |a_1(z^M)|.$$

Applying Equation (5.11) we obtain $|a_1(z^M)| = 1$, and thus Lemma 5.1.2 implies that there exists an $\alpha \in \mathbb{Z}^d$ such that $a_1(z) = \pm z^\alpha$. Consequently, due to Theorem 1.2 in [47] we obtain that the corresponding scaling function φ is of Haar–type. $\square$

5.1.2 Sum Rules

First of all, to obtain an applicable version of the sum rules (3.23), the vectors $y_\mu, |\mu| < k$, have to be determined. Fortunately, for interpolating scaling vectors, they are given explicitly.

Lemma 5.1.5. *Let $A \in \ell_0(\mathbb{Z}^d)^{m \times m}$ be the mask of an interpolating m–scaling vector Φ satisfying the sum rules of order k. Then the vectors $y_\mu, |\mu| < k$, satisfy*

$$y_\mu = [\![\Phi]\!] \left(\frac{(M^{-1}\rho_0)^\mu}{\mu!}, \ldots, \frac{(M^{-1}\rho_{m-1})^\mu}{\mu!} \right)^\top$$

with $[\![\Phi]\!] = \|\widehat{\Phi}(0)\|_2$.

Proof. As stated above, the interpolation condition (3.6) implies ℓ_p–stability. Therefore, Equation (3.25) yields

$$\frac{x^\mu}{\mu!} = \sum_{0 \leq \nu \leq \mu} \sum_{\beta \in \mathbb{Z}^d} \frac{\beta^\nu}{\nu!} y_{\mu-\nu}^\top \frac{1}{[\![\Phi]\!]} \Phi(x - \beta)$$

for all $\mu \in \mathbb{Z}_+^d$ with $|\mu| < k$. Since Φ provides accuracy of order k, it holds that $x^\mu \in S(\Phi)$ for $|\mu| < k$, and, due to the sampling property (3.8), we have

$$\frac{x^\mu}{\mu!} = \sum_{\beta \in \mathbb{Z}^d} \sum_{i=0}^{m-1} \frac{(\beta + M^{-1}\rho_i)^\mu}{\mu!} \phi_i(x - \beta).$$

Since the integer translates of Φ are linearly independent we obtain

$$\frac{1}{[\![\Phi]\!]} \sum_{0 \leq \nu \leq \mu} \frac{\beta^\nu}{\nu!} y_{\mu-\nu}^{(i)} = \frac{(\beta + M^{-1}\rho_i)^\mu}{\mu!}$$

for $i = 0, \ldots, m - 1$ where $y_\mu^{(i)}$ denotes the ith component of y_μ. Induction over $|\mu|$ yields the result. □

Though, in general, $[\![\Phi]\!]$ is unknown, Lemma 5.1.5 enables us to obtain a simple and implementable version of the sum rules (3.23). This is due to the fact that the factor $[\![\Phi]\!]$ appears on both sides of the sum rules and therefore can be omitted.

Theorem 5.1.6. *Let $A \in \ell_0(\mathbb{Z}^d)^{m \times m}$ be the mask of an interpolating m–scaling vector Φ as in (3.10). The mask A satisfies the sum rules of order k if and only if*

$$\sum_{j=0}^{m-1} \sum_{\beta \in \mathbb{Z}^d} a_{\rho + M\beta}^{(j,i)} \left(M^{-1}(\rho_j - \rho) - \beta \right)^\mu = \left(M^{-2}\rho_i \right)^\mu$$

holds for all $1 \leq i < m$, $\rho \in R$ and $\mu \in \mathbb{Z}_+^d$ with $|\mu| < k$.

Proof. Applying Lemma 5.1.5 to the ith component of the vector valued Equation (3.23) we obtain

$$
\sum_{|\nu|=|\mu|} w(\mu,\nu) [\![\Phi]\!] \frac{(M^{-1}\rho_i)^\nu}{\nu!} =
$$

$$
\sum_{0\le\nu\le\mu} (-1)^{|\nu|} \sum_{\beta\in\mathbb{Z}^d} \frac{(M^{-1}\rho+\beta)^\nu}{\nu!} \sum_{j=0}^{m-1} a^{(j,i)}_{\rho+M\beta} [\![\Phi]\!] \frac{(M^{-1}\rho_j)^{\mu-\nu}}{(\mu-\nu)!}
$$

for $0 \le i < m$. Due to (3.24) this is equivalent to

$$
\frac{(M^{-2}\rho_i)^\mu}{\mu!} = \sum_{0\le\nu\le\mu} (-1)^{|\nu|} \sum_{\beta\in\mathbb{Z}^d} \frac{(M^{-1}\rho+\beta)^\nu}{\nu!} \sum_{j=0}^{m-1} a^{(j,i)}_{\rho+M\beta} \frac{(M^{-1}\rho_j)^{\mu-\nu}}{(\mu-\nu)!}.
$$

The multivariate binomial theorem yields

$$
\left(M^{-2}\rho_i\right)^\mu = \sum_{j=0}^{m-1} \sum_{\beta\in\mathbb{Z}^d} a^{(j,i)}_{\rho+M\beta} \left(M^{-1}(\rho_j-\rho)-\beta\right)^\mu.
$$

This equation holds always true for $i = 0$, since (3.10) implies $a^{(j,0)}_\beta = \delta_{\beta,\rho_j}$. Thus, the proof is complete. $\qquad\square$

For an orthonormal interpolating scaling vector with $m = 2$ we obtain the following simplification.

Corollary 5.1.7. *If we choose* $a^{(1)}(z) = z^\rho \sum_{\beta\in\mathbb{Z}^d} (-1)^{\mathbb{1}_{[\rho]}(\beta)} a_\beta z^{-\beta}$ *in* (5.10), *then for an orthonormal interpolating 2–scaling vector the sum rules are reduced to*

$$
\left(M^{-2}\rho\right)^\mu = \sum_{\beta\in\mathbb{Z}^d} a_\beta \left(-M^{-1}\beta\right)^\mu,
$$

$$
\left(M^{-2}\rho\right)^\mu = \sum_{\beta\in\mathbb{Z}^d} a_\beta \left(M^{-1}\beta\right)^\mu (-1)^{\mathbb{1}_{[\rho]}(\beta)}
$$

with $R = \{0,\rho\}$.

Proof. For the nontrivial representative $\rho \in R$, Theorem 5.1.6 yields

$$
\left(M^{-2}\rho\right)^\mu = \sum_{\beta\in\mathbb{Z}^d} a_{M\beta+\rho} \left(-M^{-1}\rho-\beta\right)^\mu + (-1)^{\mathbb{1}_{[\rho]}(-M\beta)} a_{-M\beta}(-\beta)^\mu.
$$

Since $(-1)^{\mathbb{1}_{[\rho]}(-M\beta)} = 1$, we obtain

$$
\begin{aligned}
\left(M^{-2}\rho\right)^{\mu} &= \sum_{\beta\in\mathbb{Z}^d} a_{M\beta+\rho}\left(-M^{-1}(\rho+M\beta)\right)^{\mu} + a_{M\beta}\left(M^{-1}(M\beta)\right)^{\mu} \\
&= \sum_{\beta\in\mathbb{Z}^d} a_{\beta}\left((-1)^{\mathbb{1}_{[\rho]}(\beta)}M^{-1}\beta\right)^{\mu} \\
&= \sum_{\beta\in\mathbb{Z}^d} a_{\beta}\left(M^{-1}\beta\right)^{\mu}(-1)^{|\mu|\cdot\mathbb{1}_{[\rho]}(\beta)}.
\end{aligned}
$$

On the other hand, for the representative $0 \in R$ we have

$$
\begin{aligned}
\left(M^{-2}\rho\right)^{\mu} &= \sum_{\beta\in\mathbb{Z}^d} a_{M\beta}(-\beta)^{\mu} + (-1)^{\mathbb{1}_{[\rho]}(\rho-M\beta)}a_{\rho-M\beta}(M^{-1}\rho-\beta)^{\mu} \\
&= \sum_{\beta\in\mathbb{Z}^d} a_{M\beta}\left(-M^{-1}(M\beta)\right)^{\mu} - a_{M\beta+\rho}\left(M^{-1}(\rho+M\beta)\right)^{\mu} \\
&= \sum_{\beta\in\mathbb{Z}^d} a_{\beta}\left((-1)^{\mathbb{1}_{[0]}(\beta)}M^{-1}\beta\right)^{\mu}(-1)^{\mathbb{1}_{[\rho]}(\beta)} \\
&= \sum_{\beta\in\mathbb{Z}^d} a_{\beta}\left(M^{-1}\beta\right)^{\mu}(-1)^{(1+|\mu|)\mathbb{1}_{[\rho]}(\beta)+|\mu|}.
\end{aligned}
$$

With

$$
(-1)^{|\mu|\cdot\mathbb{1}_{[\rho]}(\beta)} = \begin{cases} (-1)^{|\mu|}, & \text{if } |\mu| \text{ is even,} \\ (-1)^{\mathbb{1}_{[\rho]}(\beta)}, & \text{else,} \end{cases}
$$

and

$$
(-1)^{(1+|\mu|)\mathbb{1}_{[\rho]}(\beta)+|\mu|} = \begin{cases} (-1)^{\mathbb{1}_{[\rho]}(\beta)}, & \text{if } |\mu| \text{ is even,} \\ (-1)^{|\mu|}, & \text{else,} \end{cases}
$$

the proof is complete. $\square$

Remark 5.1.8. *With the above choice, if $A \in \ell_0(\mathbb{Z}^d)^{2\times2}$ satisfies the sum rules of order 1, we have*

$$
\mathbf{A}(1) = \begin{pmatrix} 1 & 1 \\ 1 & 1 \end{pmatrix}
$$

and thus $\mathbf{A}(1)$ has the eigenvalues 2 and 0. Therefore, Theorem 3.1.1 ensures the existence and uniqueness of a compactly supported scaling vector corresponding to A. In addition, condition (iii) of Theorem 3.2.1 is satisfied.

5.2   Explicit Construction

In this section, we give an explicit construction method for the symbols of orthonormal interpolating 2–scaling vectors on $\mathbb{R}^d$ with compact support. Since the
construction involves necessary conditions on the mask only, we also explain how
to verify sufficient conditions. To substantiate our approach, several examples for
the case $d = 2$ are presented.

5.2.1   General Method

Based on the results in the preceding section we suggest the following construction
principle:

1. Choose a scaling matrix M with $|\det(M)| = 2$ and the nontrivial represen-
   tative ρ of $\mathbb{Z}^d/M\mathbb{Z}^d$ such that $R = \{0, \rho\}$.

2. Start with the first symbol entry

$$a^{(0)}(z) = \sum_{\beta \in \Lambda} a_\beta z^\beta$$

   by choosing the support $\Lambda \subset \mathbb{Z}^d$ of $(a_\beta)_{\beta \in \Lambda}$. It is a common observation that
   centering a mask around its zeroth coefficient leads to the most regular gen-
   erators. However, no rigorous mathematical justification for this approach
   has been found yet. Nevertheless, also in our case we observe that centering
   the coefficients around a_0 provides the best results, therefore we suggest the
   choice of $\Lambda = [-n, n]^d \cap \mathbb{Z}^d$.

3. According to Theorem 5.1.1 and Corollary 5.1.3 the second symbol entry
   $a^{(1)}(z)$ has to have the form

$$a^{(1)}(z) = \pm z^\alpha \sum_{\beta \in \Lambda} (-1)^{\mathbb{1}_{[\rho]}(\beta)} a_\beta z^{-\beta}$$

   with $\alpha \in [\rho]$. Based on our observations we suggest to choose $\alpha = \rho$ and
   a positive sign, since this seems to provide the highest regularity and the
   smallest support.

4. Apply the orthogonality condition (5.9) to the coefficient sequence $(a_\beta)_{\beta \in \Lambda}$.
   This will consume about one half of the degrees of freedom.

5. Finally, apply the sum rules of Corollary 5.1.7 up to the highest possible
   order to the coefficient sequence $(a_\beta)_{\beta \in \Lambda}$.

By this method we obtain a system of linear and quadratic equations in the variables $a := (a_\beta)_{\beta \in \Lambda}$. Hence, we have to solve a system of the form

$$F(a) = b, \tag{5.13}$$

where F is the nonlinear function given by our system of equations and b denotes the corresponding right hand side. Due to the large number of equations, it is rather impossible to solve this system analytically. Therefore, a numerical method has to be applied. Since we know F and its Jacobian F' explicitly, we suggest to formulate the problem (5.13) as a nonlinear least squares problem

$$\|F(a) - b\|_2^2 = \min. \tag{5.14}$$

This problem can be solved with the the Gauß–Newton method, i.e., starting with a vector $\tilde{a}^0 \in \mathbb{R}^{|\Lambda|}$ we have the iteration rule

$$\tilde{a}^{n+1} := \tilde{a}^n - \left(F'(\tilde{a}^n)^\top F'(\tilde{a}^n)\right)^{-1} F'(\tilde{a}^n) F(\tilde{a}^n), \tag{5.15}$$

see, e.g., [44] or another textbook on numerical analysis for details. We stop the iteration if the norm of the residual $\|F(a) - b\|_2$ becomes smaller than some tolerance $\varepsilon > 0$.

As we have to deal with quadratic equations, the solutions of this system are by no means unique. So, as a screening process, we suggest to measure/estimate the regularity of the scaling vectors corresponding to the obtained solutions. This can be performed by applying an implementation of the method described in Section 3.2.3.

Sufficient Conditions

So far, the conditions on the mask involved in our approach are necessary only. Therefore, during the construction process, one has to check whether the corresponding scaling vectors actually do possess the desired properties.

If a mask A obtained by our approach satisfies the sum rules of order 1 then, due to Remark 3.2.6, condition (ii) of Theorem 3.2.1 holds true. Furthermore, as stated in Remark 5.1.8, condition (iii) of Theorem 3.2.1 is satisfied as well. Thus, to ensure the orthonormality of the corresponding scaling vector Φ, we have to check that $\Phi \in L_2(\mathbb{R}^d)^m$ and that the associated transition operator $T_{\mathfrak{A},K}$ satisfies the eigenvalue condition (iv). These tasks can be regarded as subtasks of the regularity estimation by means of Theorem 3.2.7. There we already have to compute the spectrum of $T_{\mathfrak{A},K}$, and $\mathfrak{s}(\Phi) > 0$ implies $\Phi \in L_2(\mathbb{R}^d)^m$.

On the other hand, to ensure that a scaling vector Φ is interpolating, we have to check that Φ is continuous and satisfies $\Phi(M^{-1}\beta) = (\delta_{0,\beta}, \delta_{\rho,\beta})^\top$ for all $\beta \in \mathbb{Z}^d$.

This can be performed by utilizing the eigenvector trick introduced at the end of Section 3.1.5. First of all, if $\mathfrak{s}(\Phi) > d/2$, the Sobolev embedding theorem implies that Φ is continuous. A straightforward computation shows that $(\delta_{0,\beta}, 0)_{\beta \in \mathbb{Z}^d}^\top$ is an eigenvector of T_A corresponding to the eigenvalue 1, and thus condition (3.11) implies that Φ is interpolating.

5.2.2   The Case $d = 2$

To show the potential of our approach, several examples of interpolating 2–scaling vectors on $\mathbb{R}^2$ are constructed in the sequel. We focus on two of the most popular scaling matrices with determinant ± 2, i.e., the *quincunx matrices* M_q and M_p, defined by

$$M_q := \begin{pmatrix} 1 & -1 \\ 1 & 1 \end{pmatrix} \quad \text{and} \quad M_p := \begin{pmatrix} 1 & 1 \\ 1 & -1 \end{pmatrix}.$$

Both matrices are idempotent and generate the commonly known *quincunx lattice*, i.e., $M_q \mathbb{Z}^2 = M_p \mathbb{Z}^2 = \{(i,j)^\top \in \mathbb{Z}^2 \mid i+j \text{ is even}\}$. Therefore, the cosets of $\mathbb{Z}^2/M_q\mathbb{Z}^2$ and $\mathbb{Z}^2/M_p\mathbb{Z}^2$ coincide and we choose $\rho := (0,1)^\top$ in both cases. In contrast to M_q, the matrix M_p satisfies

$$M^2 = 2\mathbf{I}_d \tag{5.16}$$

with $d = 2$. Usually, scaling matrices satisfying (5.16) are also called *box spline matrices*. A *box spline* is a refinable function defined on $\mathbb{R}^d$ which can be considered as a multivariate generalization of classical cardinal B–splines, see [41] for details. In general, box splines are refinable with respect to dyadic scaling. However, it has been shown in [29] that for each scaling matrix M satisfying (5.16) there are box splines which are refinable with respect to M as well.

   The next theorem shows that for both matrices the solutions of our equation systems are closely related.

Theorem 5.2.1. *The sequence $a^q \in \ell_0(\mathbb{Z}^2)$ satisfies the orthogonality condition (5.9) and the sum rules of order k in Corollary 5.1.7 with respect to $M = M_q$ if and only if the sequence $a^p \in \ell_0(\mathbb{Z}^2)$, defined by*

$$a_\beta^p := (-1)^{\mathbf{1}_{[\rho]}(\beta)} a_{U\beta}^q \quad \text{with} \quad U := \begin{pmatrix} -1 & 0 \\ 0 & 1 \end{pmatrix},$$

satisfies the corresponding conditions with respect to $M = M_p$.

Proof. Let us show the orthogonality condition first. For $\gamma \in \mathbb{Z}^2$ it holds that

$$\sum_{\beta \in \mathbb{Z}^2} a_\beta^p a_{\beta - M_p\gamma}^p = \sum_{\beta \in \mathbb{Z}^2} a_{U\beta}^q a_{U(\beta - M_p\gamma)}^q = \sum_{\beta \in \mathbb{Z}^2} a_\beta^q a_{\beta - UM_p\gamma}^q.$$

Since $UM_p\mathbb{Z}^2 = M_p\mathbb{Z}^2 = M_q\mathbb{Z}^2$, there exists a $\tilde{\gamma} \in \mathbb{Z}^2$ with $UM_p\gamma = M_q\tilde{\gamma}$. Therefore, we have

$$\sum_{\beta \in \mathbb{Z}^2} a_\beta^p a_{\beta - M_p\gamma}^p = \sum_{\beta \in \mathbb{Z}^2} a_\beta^q a_{\beta - M_q\tilde{\gamma}}^q$$

and with $\delta_{0,\tilde{\gamma}} = \delta_{0,\gamma}$ the orthogonality conditions (5.9) for a^p and for a^q are equivalent.

On the other hand, for $\mu \in \mathbb{Z}_+^2$ with $|\mu| < k$ we have

$$\begin{aligned}
\sum_{\beta \in \mathbb{Z}^2} a_\beta^p \left(-M_p^{-1}\beta\right)^\mu &= \sum_{\beta \in \mathbb{Z}^2} a_{U\beta}^q \left(-M_p^{-1}\beta\right)^\mu (-1)^{\mathbb{1}_{[\rho]}(\beta)} \\
&= \sum_{\beta \in \mathbb{Z}^2} a_\beta^q \left(-M_p^{-1}U\beta\right)^\mu (-1)^{\mathbb{1}_{[\rho]}(U\beta)}.
\end{aligned}$$

Using the notation $\mu := (\mu_0, \mu_1)^\top$ it holds that

$$\left(-M_p^{-1}U\beta\right)^\mu = \left(M_q^{-\top}\beta\right)^\mu = (-1)^{\mu_0}\left(M_q^{-1}\beta\right)^{E\mu}$$

with $E := \begin{pmatrix} 0 & 1 \\ 1 & 0 \end{pmatrix}$. With $\mathbb{1}_{[\rho]}(U\beta) = \mathbb{1}_{[\rho]}(\beta)$, this leads to

$$\sum_{\beta \in \mathbb{Z}^2} a_\beta^p \left(-M_p^{-1}\beta\right)^\mu = (-1)^{\mu_0} \sum_{\beta \in \mathbb{Z}^2} a_\beta^q \left(M_q^{-1}\beta\right)^{E\mu} (-1)^{\mathbb{1}_{[\rho]}(\beta)}.$$

Furthermore, it holds that

$$\left(M_p^{-2}\rho\right)^\mu = (-1)^{\mu_0}\left(M_q^{-2}\rho\right)^{E\mu}.$$

Therefore, due to $|\mu| = |E\mu|$, the first part of the sum rules of order k in Corollary 5.1.7 for a^p is equivalent to the second part of the sum rules of order k for a^q. The opposite direction can be shown analogously. $\qquad\square$

Numerical Issues

In the numerical treatment of our equation system we are confronted with two major problems. The first problem is related to the local convergence of the Gauß–Newton method. We have to select a sufficiently large set of starting vectors $\tilde{a}^0 \in \mathbb{R}^{|\Lambda|}$ in order to find at least some good solutions. This can be performed by either uniformly or randomly sampling $\mathbb{R}^{|\Lambda|}$ (or the unit sphere in $\mathbb{R}^{|\Lambda|}$, as Equation (5.9) implies $\|(a_\beta)\|_2 = 1$). In both cases, for Λ becoming large, we have to deal with a rapidly increasing amount of starting vectors. For the scaling matrix M_q, this difficulty can be eased by the observation that many solutions which correspond to scaling vectors with a high regularity share the structure

$$a_\beta = \begin{cases} a_{E\beta}, & \text{if } \beta \in M_q \mathbb{Z}^2, \\ -a_{-E\beta}, & \text{else}, \end{cases} \quad \text{with} \quad E = \begin{pmatrix} 0 & 1 \\ 1 & 0 \end{pmatrix}. \tag{5.17}$$

According to Theorem 5.2.1, a similar structure can be found for M_p. For large index sets Λ, we focus on these specific structures.

The second problem is to distinguish true solutions from local minima. It turns out that even if the norm of the residual is close to *machine accuracy*, i.e., the smallest floating point number ε_0 on a computer such that $1 \pm \varepsilon_0 \neq 1$, the corresponding solution may still belong to a local minimum. Therefore, we test our solutions with a multiple precision implementation of the Gauß–Newton method. If the norm of the residual can be reduced significantly below ε_0, then we assume the corresponding solution to be at least reasonable. Obviously, it is rather unlikely that the multiple precision algorithm ends up with a true solution, thus we have to choose a lower bound ε for the norm of the residual as a stopping criterion. In practice, most computers work with standard 64 bit floating point arithmetic where $\varepsilon_0 \approx 2.22 \cdot 10^{-16}$, cf. [3]. Hence, $\varepsilon := 10^{-25}$ seems to be an appropriate choice for the stopping parameter. Nevertheless, for most applications even a larger norm of the residual may be perfectly fine.

5.2.3   Examples

Starting with an index set $\Lambda = \{-n, \dots, n\}^2$ we obtain a sequence of scaling vectors denoted by Φ_n with increasing accuracy order and regularity. It turns out that for both scaling matrices, those solutions of our equation systems which correspond to the scaling vectors with the highest regularity are linked via Theorem 5.2.1. The corresponding scaling vectors shall be denoted by Φ_n^q for dilation with M_q and Φ_n^p for dilation with M_p. In Table 5.1 the properties of the constructed examples are shown, the corresponding masks can be found in Appendix A.2.

For the case $n = 0$ our solutions are the characteristic functions of the multi–tiles shown in Figure 5.1. Via the rule (3.7), these scaling vectors coincide with the characteristic functions of the classical tiles corresponding to the scaling matrices M_q and M_p, cf. [53]. Both symbols have the form

$$\mathbf{A}(z) = \begin{pmatrix} 1 & 1 \\ z^\rho & z^\rho \end{pmatrix}.$$

For $n \geq 2$ all our solutions have critical Sobolev exponents strictly larger than one. Therefore, by the Sobolev embedding theorem, all these scaling vectors are at

n	accuracy order	$\mathfrak{s}(\Phi_n)$	$\|\text{residual}\|_\infty$
$M = M_q$			
0	1	0.238	$< \varepsilon$
1	1	0.743	$< \varepsilon$
2	2	1.355	$< \varepsilon$
3	3	1.699	$< \varepsilon$
4	3	1.819	$< 10^{-18}$
5	4	2.002	$< 10^{-22}$
$M = M_p$			
0	1	0.5	$< \varepsilon$
1	1	0.736	$< \varepsilon$
2	2	1.371	$< \varepsilon$
3	3	1.695	$< \varepsilon$
4	3	1.934	$< 10^{-18}$
5	4	2.099	$< 10^{-22}$

Table 5.1: Properties of the Φ_n

least continuous. Figures 5.2 and 5.3 show the component functions of Φ_2^q and Φ_2^p, respectively. Furthermore, for $n = 5$ we obtain an example that is continuously differentiable. The corresponding functions are graphed in Figures 5.4 and 5.5. The reader should note that all these scaling vectors are very well localized.

5.3   Multiwavelets

In this section, we will show that also in the multivariate case, our interpolating scaling vectors lead to (orthonormal) multiwavelet bases in a very simple and natural way. We focus on the case $r = m = 2$. Thus, according to Section 3.1.3, we only have to deal with one multiwavelet $\Psi := (\psi_0, \psi_1)^\top$ with symbol $\mathbf{B}(z)$.

Let Φ be a compactly supported orthonormal 2–scaling vector with symbol $\mathbf{A}(z)$ which consequently generates an MRA. As we have seen in Section 3.2.1, the task of finding a multiwavelet Ψ can be converted into a matrix extension problem. Thus, we have to find $\mathbf{B}(z)$ such that $\mathcal{P}_m(z)$ defined in (3.20) is unitary for all $z \in \mathbb{T}^d$. Similar to the univariate case, the solution of this problem is by no means unique. However, if Φ is interpolating and Ψ is interpolating as well, i.e., we have

$$\Psi\left(M^{-1}\beta\right) = \begin{pmatrix} \delta_{0,\beta} \\ \delta_{\rho,\beta} \end{pmatrix} \quad \text{for all} \quad \beta \in \mathbb{Z}^d$$

and for $R = \{0, \rho\}$, then the extension problem has a unique solution of the following form.

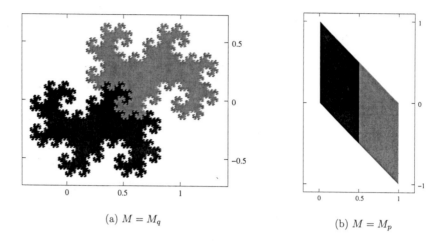

(a) $M = M_q$ (b) $M = M_p$

Figure 5.1: (Multi-) tiles corresponding to Φ_0

Theorem 5.3.1. *Let $\mathbf{A}(z)$ be the symbol of a compactly supported orthonormal interpolating 2–scaling vector Φ. Furthermore, let $\mathbf{B}(z)$ be the symbol of an interpolating function vector Ψ defined by (3.12). The matrix $\mathcal{P}_m(z)$ in (3.20) is unitary for all $z \in \mathbb{T}^d$ if and only if*

$$\mathbf{B}(z) = \begin{pmatrix} 1 & -a^{(0)}(z) \\ z^\rho & -a^{(1)}(z) \end{pmatrix} \tag{5.18}$$

holds with $a^{(0)}(z)$ and $a^{(1)}(z)$ as in (3.11).

Proof. Since Φ and Ψ are interpolating, a direct computation using (3.12) yields that the symbol $\mathbf{B}(z)$ has to have the form

$$\mathbf{B}(z) = \begin{pmatrix} 1 & b^{(0)}(z) \\ z^\rho & b^{(1)}(z) \end{pmatrix}$$

for some Laurent polynomials $b^{(0)}(z)$ and $b^{(1)}(z)$. Let $\mathcal{P}_m(z)$ be unitary, then it holds that

$$\mathbf{A}(z)\overline{\mathbf{B}(z)}^\top + \mathbf{A}(z_{M^{-\top}\tilde{\rho}})\overline{\mathbf{B}(z_{M^{-\top}\tilde{\rho}})}^\top = \mathbf{0}. \tag{5.19}$$

Thus, if we define $\mathcal{A}(z)$ as in (5.6) and

$$\mathcal{B}(z) := \frac{1}{\sqrt{2}} \begin{pmatrix} b^{(0)}(z) & b^{(0)}(z_{M^{-\top}\tilde{\rho}}) \\ b^{(1)}(z) & b^{(1)}(z_{M^{-\top}\tilde{\rho}}) \end{pmatrix},$$

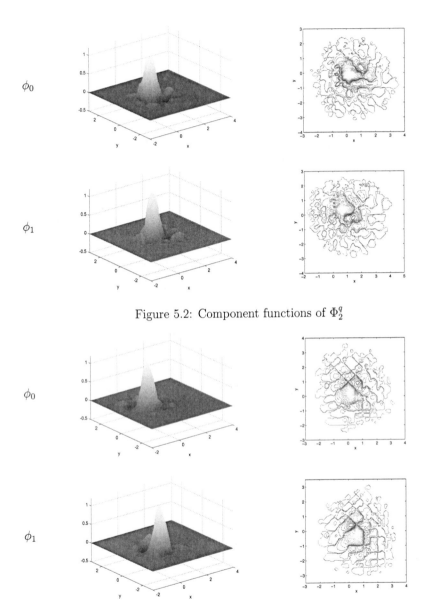

Figure 5.2: Component functions of Φ_2^q

Figure 5.3: Component functions of Φ_2^p

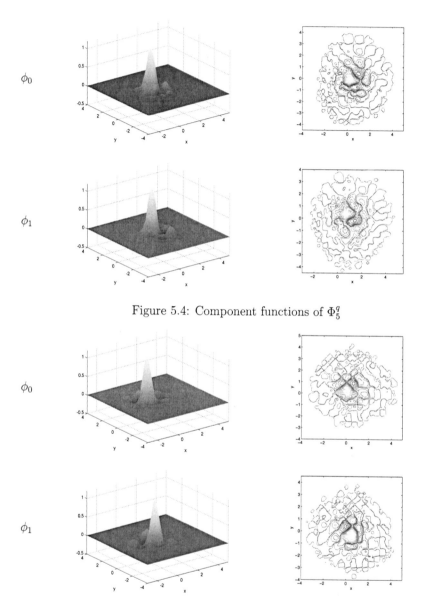

Figure 5.4: Component functions of Φ_5^q

Figure 5.5: Component functions of Φ_5^p

then, with $z^\alpha + z^\alpha_{M^{-\top}\tilde{\rho}} = 0$ for $\alpha \in [\rho]$, Equation (5.19) is equivalent to

$$\mathcal{A}(z)\overline{\mathcal{B}(z)}^{\top} = -\mathbf{I}_2. \tag{5.20}$$

Due to Theorem 5.1.1, we have $a^{(1)}(z) = \pm z^\alpha \overline{a^{(0)}(z_{M^{-\top}\tilde{\rho}})}$, $\alpha \in [\rho]$, and it holds that $\det(\mathcal{A}(z)) = \mp z^\alpha$. Applying Cramer's rule to (5.20), we obtain

$$\overline{b^{(0)}(z)} = \frac{-a^{(1)}(z_{M^{-\top}\tilde{\rho}})}{\mp z^\alpha} = \frac{z^\alpha_{M^{-\top}\tilde{\rho}}\overline{a^{(0)}(z)}}{z^\alpha} = -\overline{a^{(0)}(z)}$$

and

$$\overline{b^{(1)}(z)} = \frac{a^{(0)}(z_{M^{-\top}\tilde{\rho}})}{\mp z^\alpha} = \frac{\pm z^\alpha \overline{a^{(1)}(z)}}{\mp z^\alpha} = -\overline{a^{(1)}(z)}.$$

Thus $\mathbf{B}(z)$ has to have of the form (5.18).

On the other hand, if a symbol $\mathbf{B}(z)$ corresponding to a mask $(B_\beta) \in \ell_0(\mathbb{Z}^d)^{2\times 2}$ satisfies (5.18), it is easy to verify that $\mathcal{P}_m(z)$ is unitary for all $z \in \mathbb{T}$. $\square$

This theorem provides us with a convenient method to construct a multiwavelet Ψ corresponding to an orthonormal interpolating scaling vector Φ with compact support. Since the components of Φ are not normalized, the same holds for the components of Ψ. However, Theorem 3.2.3 implies that $\sqrt{2}\Phi$ is strictly orthonormal, and thus we obtain the following corollary.

Corollary 5.3.2. *Under the assumptions of Theorem 5.3.1 let (5.18) be satisfied. Then $\sqrt{2}\Psi$ gives rise to an orthonormal multiwavelet basis.*

Though the component functions of Ψ in 5.3.1 are not normalized, Ψ may be called an *orthonormal interpolating multiwavelet*. Some examples of such interpolating multiwavelets are shown in Figures 5.6 and 5.7. These examples correspond to our scaling vectors Φ^q_n and Φ^p_n for $n = 2$ and $n = 5$ and are denoted by Ψ^q_n and Ψ^p_n, respectively.

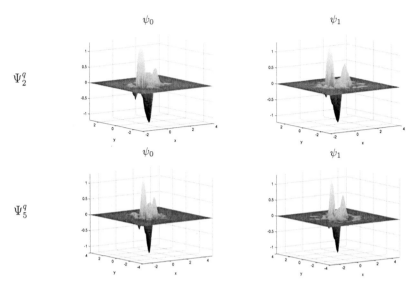

Figure 5.6: Multiwavelets corresponding to Φ^q

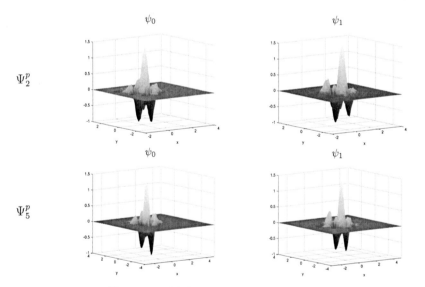

Figure 5.7: Multiwavelets corresponding to Φ^p

Chapter 6

Recipe III: Multivariate Symmetric Interpolating Scaling Vectors

In this chapter, our aim is to extend our approach obtained in the previous chapter by incorporating an additional property: symmetry. That means, we intend to construct interpolating scaling vectors which are symmetric, possess good approximation and regularity properties, and lead to nice multiwavelet bases simultaneously. Although we cannot exclude that there may exist multivariate orthonormal interpolating scaling vectors which are symmetric and possess compact support, the task of constructing such scaling vectors seems to be somewhat unpromising. We have seen in the preceding chapters that interpolation plus orthonormality does only leave a narrow margin for incorporating further properties. So, even if we manage to construct such all-in-one scaling vectors, to provide reasonable approximation power their support is likely to grow horribly large. Therefore, we focus on the biorthogonal case which provides more flexibility.

Similar to the approaches in Chapters 4 and 5, we start our construction by collecting necessary conditions for the desired properties in terms of the mask or symbol, respectively. The results obtained in this chapter have already been published in [81].

6.1 Main Ingredients

The main benefit of interpolating scaling vectors is the sampling property (3.8) which provides us with a convenient method for obtaining the coefficients of a function in $S(\Phi)$. In general, when applying biorthogonal scaling vectors or functions, one only needs the representation of a function either in terms of the primal or in terms of the dual generator. Then the corresponding dual pair of filter banks is used to analyze and synthesize the coefficient sequence. Therefore, we focus on interpolating scaling vectors on the primal side. Their duals do not necessarily have to share this property.

6.1.1   Symmetry

In contrast to the univariate case, multivariate functions can possess various symmetries. Moreover, since a scaling matrix M can feature several rotation and reflection properties, the characteristics of M have to be taken into account when constructing symmetric scaling functions or vectors. The following notion of symmetry was stated in [60], see also [57].

A finite set $\mathcal{G} \subset \{U \in \mathbb{Z}^{d \times d} \,|\, |\det U| = 1\}$ is called a *symmetry group with respect to* M if $\mathcal{G}$ forms a group under matrix multiplication and for all $U \in \mathcal{G}$ we have $MUM^{-1} \in \mathcal{G}$. Since $\mathcal{G}$ is finite, $U \in \mathcal{G}$ implies $M^{-1}UM \in \mathcal{G}$ as well. A function $f : \mathbb{R}^d \to \mathbb{R}^d$ is called $\mathcal{G}$*-symmetric with center* $c_f \in \mathbb{R}^d$ if for all $U \in \mathcal{G}$ and for all $x \in \mathbb{R}^d$ we have

$$f(U(x - c_f) + c_f) = f(x),$$

i.e., for each point $x \in \mathbb{R}^d$, shifting the whole space with the center c_f towards the origin, rotating or reflecting by means of U, and shifting back preserves the value of f. A simple example of such a symmetry group is $\mathcal{G} = \{\pm \mathbf{I}_d\}$ which, with $c_f := 0$, resembles the classical univariate notion of symmetry. However, in the multivariate setting one is often interested in more complex symmetries and thus larger symmetry groups. Since all elements of $\mathcal{G}$ are integer matrices, the notion of symmetry can be used for sequences as well. Consequently, a sequence $(a_\beta)_{\beta \in \mathbb{Z}^d}$ is called $\mathcal{G}$*-symmetric with center* $c_a \in \mathbb{R}^d$ if

$$a_{U(\beta - c_a) + c_a} = a_\beta$$

holds for all $U \in \mathcal{G}$ and for all $\beta \in \mathbb{Z}^d$. Clearly, one has to ensure that for all $U \in \mathcal{G}$ it holds that $U(\beta - c_a) + c_a \in \mathbb{Z}^d$.

The requirement $MUM^{-1} \in \mathcal{G}$ aims at connecting the symmetry of a refinable function (vector) with the symmetry of its mask. It is commonly known that in the scalar case the symmetry properties of the scaling vector and those of its mask are closely related, cf. [60]. In the vector setting, due to the more complex algebraic structure of the mask, this relation is somewhat more involved.

Theorem 6.1.1. *Let* $\mathcal{G}$ *be a symmetry group with respect to* M *and let* $c_i \in \mathbb{R}^d$, $0 \le i < r$, *such that* $Uc_i - c_i \in \mathbb{Z}^d$ *for all* $U \in \mathcal{G}$. *Furthermore, let* Φ *be an* r-*scaling vector with mask* $A \in \ell_0(\mathbb{Z}^d)^{r \times r}$ *as in* (3.2). *If the component functions* ϕ_i

of Φ are $\mathcal{G}$–symmetric with centers c_i, then the $(a_\beta^{(i,j)})_{\beta\in\mathbb{Z}^d}$ are $\mathcal{G}$–symmetric with centers $Mc_i - c_j$.

Proof. For $0 \leq i < r$ and $x \in \mathbb{R}^d$ the refinement equation implies

$$\phi_i(x) = \sum_{j=0}^{r-1} \sum_{\beta\in\mathbb{Z}^d} a_\beta^{(i,j)} \phi_j(Mx - \beta).$$

For an arbitrary $U \in \mathcal{G}$ the symmetry of the ϕ_j yields

$$\phi_i(x) = \sum_{j=0}^{r-1} \sum_{\beta\in\mathbb{Z}^d} a_\beta^{(i,j)} \phi_j(UMx - U\beta - Uc_j + c_j).$$

Since $U \in \mathcal{G}$, there exists a $V \in \mathcal{G}$ such that $UM = MV$, and with $Uc_j - c_j \in \mathbb{Z}^d$ we obtain

$$\phi_i(x) = \sum_{j=0}^{r-1} \sum_{\beta\in\mathbb{Z}^d} a_{U^{-1}(\beta+c_j)-c_j}^{(i,j)} \phi_j(MVx - \beta).$$

On the other hand, we have

$$\phi_i(x) = \phi_i(V(x - c_i) + c_i) \;=\; \sum_{\beta\in\mathbb{Z}^d}\sum_{j=0}^{r-1} a_\beta^{(i,j)} \phi_j(MV(x - c_i) + Mc_i - \beta)$$

$$= \sum_{\beta\in\mathbb{Z}^d}\sum_{j=0}^{r-1} a_{\beta-MVc_i+Mc_i}^{(i,j)} \phi_j(MVx - \beta)$$

since $Vc_i - c_i \in \mathbb{Z}^d$. Thus, with $UM = MV$ we obtain

$$a_\beta^{(i,j)} = a_{U(\beta-Mc_i+c_j)+Mc_i-c_j}^{(i,j)}.$$

$\square$

Remark 6.1.2. *Assume that an interpolating scaling vector Φ is $\mathcal{G}$–symmetric with centers c_i, $0 \leq i < m$. Then, the interpolation condition (3.6) implies for all $U \in \mathcal{G}$ and $0 \leq i < m$*

$$1 = \phi_i(M^{-1}\rho_i) = \phi_i(UM^{-1}\rho_i - Uc_i + c_i).$$

Since there exists a $V \in \mathcal{G}$ with $UM^{-1} = M^{-1}V$, it follows that

$$1 = \phi_i(M^{-1}(V\rho_i - VMc_i + Mc_i)) = \delta_{\rho_i,V\rho_i-VMc_i+Mc_i},$$

and consequently $V(\rho_i - Mc_i) = \rho_i - Mc_i$. Hence, we either have $c_i = M^{-1}\rho_i$ for all $0 \leq i < m$, or all symmetry matrices $V \in \mathcal{G}$ have at least one eigenvector corresponding to the eigenvalue 1 in common. The latter case, though being not impossible, seems to be rather artificial. Therefore, we focus on the first case, i.e., Φ is called $\mathcal{G}$–symmetric if we have $c_i = M^{-1}\rho_i$ for $0 \leq i < m$.

Corollary 6.1.3. *Let $\mathcal{G}$ be a symmetry group with respect to M and let Φ be a $\mathcal{G}$–symmetric interpolating m–scaling vector with mask $A \in \ell_0(\mathbb{Z}^d)^{r \times r}$. If $U[\rho] = [\rho]$ holds for all $\rho \in R$ and for all $U \in \mathcal{G}$, then the mask entries $(a_\beta^{(i,j)})_{\beta \in \mathbb{Z}^d}$ are $\mathcal{G}$–symmetric with centers $\rho_i - M^{-1}\rho_j$.*

Proof. The component functions ϕ_i of Φ are $\mathcal{G}$–symmetric with centers $c_i :=$ $M^{-1}\rho_i$. It remains to be shown that $Uc_i - c_i \in \mathbb{Z}^d$ holds for all $U \in \mathcal{G}$. By definition of $\mathcal{G}$, for an arbitrary $U \in \mathcal{G}$ there exists a $V \in \mathcal{G}$ such that $V = MUM^{-1}$. Furthermore, for each $\alpha \in [\rho_i]$ there exists a $\beta \in \mathbb{Z}^d$ such that $\alpha = M\beta + \rho_i$, and hence we have

$$V\alpha = VM\beta + V\rho_i = MU\beta + V\rho_i.$$

Therefore, $V[\rho_i] = [\rho_i]$ is equivalent to $V\rho_i \in [\rho_i]$ which implies that there exists a $\beta' \in \mathbb{Z}^d$ such that $V\rho_i = M\beta' + \rho_i$ or, equivalently, $V\rho_i - \rho_i = M\beta' \in M\mathbb{Z}^d$. Thus, we have $MUM^{-1}\rho_i - \rho_i \in M\mathbb{Z}^d$ which is equivalent to $UM^{-1}\rho_i - M^{-1}\rho_i \in \mathbb{Z}^d$. $\quad\square$

Remark 6.1.4. *For the case $m = |\det(M)| = 2$ the requirement $U[\rho] = [\rho]$, $\rho \in \{0, \rho_1\}$, is met for all symmetry groups $\mathcal{G}$. Obviously, $U[0] = [0]$ for all $U \in \mathcal{G}$. Let us assume that also $U\rho_1 \in [0]$ for some $U \in \mathcal{G}$. Hence, there exists a $\beta \in \mathbb{Z}^d$ such that $U\rho_1 = M\beta$. By definition of $\mathcal{G}$, $U^{-1} \in \mathcal{G}$ and there is a $V \in \mathcal{G}$ such that $U^{-1}M = MV$. It follows that $\rho_1 = U^{-1}M\beta = MV\beta \in [0]$. This contradiction yields $U[\rho_1] = [\rho_1]$.*

Theorem 6.1.1 shows that a symmetric interpolating scaling vector is completely determined by a small part of its mask. To exploit this redundancy in our construction, we have to investigate the properties of finite sets which are invariant under the action of $\mathcal{G}$. The following results are standard results from the theory of permutation groups, see, e.g., [13] for details.

A finite set $\Omega \subset \mathbb{R}^d$ is called $\mathcal{G}$–symmetric or a $\mathcal{G}$–space if for all $U \in \mathcal{G}$ it holds that $U\Omega \subset \Omega$. Then, $\mathcal{G}$ is a permutation group on Ω via the group action

$$\mathcal{G} \times \Omega \ni (U, x) \to Ux \in \Omega.$$

For $x \in \Omega$ the set $\mathcal{G}x := \{Ux \,|\, U \in \mathcal{G}\}$ is called the *orbit* of x. Since two orbits are either disjoint or equal, one can show that Ω can be decomposed into a disjoint union of orbits.

Lemma 6.1.5. *For each $\mathcal{G}$–space $\Omega \subset \mathbb{R}^d$ there exists a subset $\Lambda \subset \Omega$ such that*

$$\Omega = \bigcup_{x \in \Lambda} \mathcal{G}x,$$

where the orbits $\mathcal{G}x$, $x \in \Lambda$, are mutually disjoint.

The subgroup $\mathcal{G}_x := \{U \in \mathcal{G} \mid Ux = x\}$ of $\mathcal{G}$ is called the *stabilizer* of x. It is well known that for $x \in \Omega$ the sets $\mathcal{G}x$ and $\mathcal{G}/\mathcal{G}_x$ are isomorphic. Let G_x denote a complete set of representatives of the cosets of $\mathcal{G}/\mathcal{G}_x$, then we have the following lemma.

Lemma 6.1.6. *For $x \in \Omega$ it holds that $\mathcal{G}x = \bigcup_{U \in G_x} \{Ux\}$.*

These lemmata enable us to find a decomposition of the support of our masks.

Proposition 6.1.7. *Let $a^{(i,j)} \in \ell_0(\mathbb{Z}^d)$, $0 \le i, j < m$, be $\mathcal{G}$–symmetric with centers $c(i,j) := \rho_i - M^{-1}\rho_j$. Then there exist finite sets $\Omega_j \subset \mathbb{Z}^d$ such that $\Omega_j + M^{-1}\rho_j$ is $\mathcal{G}$–symmetric and $\mathrm{supp}(a^{(i,j)}) \subset \Omega_j + \rho_i$. Furthermore, there exist sets $\Lambda_j \subset \Omega_j$ such that we have the disjoint decomposition*

$$\Omega_j + \rho_i = \bigcup_{\beta \in \Lambda_j + \rho_i} \bigcup_{U \in G_{\beta - c(i,j)}} \{U(\beta - c(i,j)) + c(i,j)\}. \tag{6.1}$$

Proof. Since each $a^{(i,j)}$ is $\mathcal{G}$–symmetric with center $c(i,j) = \rho_i - M^{-1}\rho_j$, the set $\mathrm{supp}(a^{(i,j)}) - c(i,j)$ is a $\mathcal{G}$–space. Therefore, the set

$$\widetilde{\Omega}_j := \bigcup_{i=0}^{m-1} \left(\mathrm{supp}(a^{(i,j)}) - c(i,j) \right) \subset \mathbb{Z}^d + M^{-1}\rho_j$$

also is a $\mathcal{G}$–space. Thus, by Lemma 6.1.5 and Lemma 6.1.6 there exists an $\widetilde{\Lambda}_j \subset \widetilde{\Omega}_j$ such that the disjoint decomposition

$$\widetilde{\Omega}_j = \bigcup_{x \in \widetilde{\Lambda}_j} \bigcup_{U \in G_x} \{Ux\}$$

holds. Choosing $\Omega_j := \widetilde{\Omega}_j - M^{-1}\rho_j$ and $\Lambda_j := \widetilde{\Lambda}_j - M^{-1}\rho_j$ completes the proof. $\square$

6.1.2 Biorthogonality

Let $(\Phi, \widetilde{\Phi})$ be a pair of biorthogonal m–scaling vectors with respect to M. If Φ is interpolating then, similar to Theorem 5.1.1, the biorthogonality condition (3.18) is considerably simplified.

Proposition 6.1.8. *Let $(\Phi, \widetilde{\Phi})$ be a pair of dual m–scaling vectors with masks $(A_\beta), (\widetilde{A}_\beta) \in \ell_0(\mathbb{Z}^d)^{m \times m}$. If Φ is interpolating, then the biorthogonality condition (3.18) holds if and only if*

$$\widetilde{a}^{(j,0)}_{\rho_i - M\alpha} + \sum_{n=1}^{m-1} \sum_{\beta \in \mathbb{Z}^d} a^{(i,n)}_\beta \widetilde{a}^{(j,n)}_{\beta - M\alpha} = m \cdot \delta_{0,\alpha}\delta_{i,j}, \quad 0 \le i, j < m, \tag{6.2}$$

holds for all $\alpha \in \mathbb{Z}^d$.

Proof. For $0 \leq i, j < m$, one component of (3.18) is equivalent to

$$
\begin{aligned}
m^2 \delta_{i,j} &= \sum_{\widetilde{\rho} \in \widetilde{R}} \sum_{n=0}^{m-1} a_{i,n}(z_{M^{-\top} \widetilde{\rho}}) \overline{\widetilde{a}}_{j,n}(z_{M^{-\top} \widetilde{\rho}}) \\
&= \sum_{\widetilde{\rho} \in \widetilde{R}} \sum_{n=0}^{m-1} \sum_{\alpha, \beta \in \mathbb{Z}^d} a_{\beta}^{(i,n)} \widetilde{a}_{\beta-\alpha}^{(j,n)} z_{M^{-\top} \widetilde{\rho}}^{\alpha} \\
&= \sum_{n=0}^{m-1} \sum_{\alpha, \beta \in \mathbb{Z}^d} \sum_{\rho \in R} a_{\beta}^{(i,n)} \widetilde{a}_{\beta-(M\alpha+\rho)}^{(j,n)} \sum_{\widetilde{\rho} \in \widetilde{R}} z_{M^{-\top} \widetilde{\rho}}^{M\alpha+\rho}.
\end{aligned}
$$

For $\rho \in R$, Lemma 3.1.2 implies that

$$
\sum_{\widetilde{\rho} \in \widetilde{R}} z_{M^{-\top} \widetilde{\rho}}^{M\alpha+\rho} = m \cdot \delta_{0,\rho} z^{M\alpha}.
$$

Hence, we obtain

$$
m \cdot \delta_{i,j} = \sum_{n=0}^{m-1} \sum_{\alpha, \beta \in \mathbb{Z}^d} a_{\beta}^{(i,n)} \widetilde{a}_{\beta-M\alpha}^{(j,n)} z^{M\alpha}
$$

which is equivalent to

$$
m \cdot \delta_{i,j} \delta_{0,\alpha} = \sum_{n=0}^{m-1} \sum_{\beta \in \mathbb{Z}^d} a_{\beta}^{(i,n)} \widetilde{a}_{\beta-M\alpha}^{(j,n)}, \quad \alpha \in \mathbb{Z}^d.
$$

Applying the interpolation condition (3.10) completes the proof.   □

Thus, given the mask of a primal interpolating scaling vector, the biorthogonality condition leads to simple linear conditions on the dual mask.

6.1.3   Sum Rules

Similar to the orthonormal case, before we can incorporate the sum rules (3.23) into our construction, the vectors y_μ have to be determined. Since we assume that the primal scaling vector Φ is interpolating, we can apply Lemma 5.1.5 and obtain

$$
y_\mu = [\![\Phi]\!] \left(\frac{(M^{-1}\rho_0)^\mu}{\mu!}, \dots, \frac{(M^{-1}\rho_{r-1})^\mu}{\mu!} \right)^\top. \tag{6.3}
$$

On the other hand, given a primal scaling vector Φ, the vectors $\widetilde{y}_\mu$ corresponding to a stable dual scaling vector $\widetilde{\Phi}$ can be obtained by means of Equation (3.25). The biorthogonality condition (3.17) yields

$$
\widetilde{y}_\mu = \frac{\widetilde{c}}{\mu!} \left(\langle x^\mu, \Phi_0(x) \rangle, \dots, \langle x^\mu, \Phi_{m-1}(x) \rangle \right)^\top =: \widetilde{c} \left\langle \frac{x^\mu}{\mu!}, \Phi(x) \right\rangle, \tag{6.4}
$$

where the constant $\tilde{c}$ depends on c in (3.17) and $[\![\tilde{\Phi}]\!]$. Thus, the $\tilde{y}_\mu$ are determined by the moments of Φ up to a constant. In general, Φ is not given explicitly, therefore we have to address the problem of computing the moments of Φ. For the scalar case, it has been shown in [33] that the moments of a scaling function can be computed iteratively. Due to the more complex algebraic structure of the vector setting, the generalization of this method is somewhat more involved.

Theorem 6.1.9. *Let Φ be an r-scaling vector with mask $A \in \ell_0(\mathbb{Z}^d)^{r\times r}$ and let $\tilde{\Phi}$ be a stable dual r-scaling vector which satisfies the sum rules of order k. For $0 < n < k$ let us define $\{\nu_1, \ldots, \nu_N\} := \{\mu \in \mathbb{Z}_+^d \mid |\mu| = n\}$ and*

$$
\mathbf{W}_N := \begin{pmatrix} w(\nu_1, \nu_1) & \cdots & w(\nu_1, \nu_N) \\ \vdots & \ddots & \vdots \\ w(\nu_N, \nu_1) & \cdots & w(\nu_N, \nu_N) \end{pmatrix}.
$$

Moreover, for $1 \leq i \leq N$ we use the notation

$$
\tilde{v}_A(\nu_i) := \sum_{0<\kappa\leq\nu_i} J_A(\kappa)\tilde{y}_{\nu_i-\kappa} \quad \text{with} \quad J_A(\kappa) := \frac{1}{m}\sum_{\beta\in\mathbb{Z}^d} \frac{\beta^\kappa}{\kappa!} A_\beta.
$$

Then, with $\mathbf{P}_N := \mathbf{I}_{rN} - \mathbf{W}_N \otimes J_A(0)$ and $\mathbf{Q}_N := \mathbf{W}_N \otimes \mathbf{I}_r$, the $\tilde{y}_\mu$ satisfy the recursion

$$
\begin{pmatrix} \tilde{y}_{\nu_1} \\ \vdots \\ \tilde{y}_{\nu_N} \end{pmatrix} = \mathbf{P}_N^{-1}\mathbf{Q}_N \begin{pmatrix} \tilde{v}_A(\nu_1) \\ \vdots \\ \tilde{v}_A(\nu_N) \end{pmatrix}.
$$

Proof. For $1 \leq i \leq N$, applying the refinement equation to (6.4) we obtain

$$
\tilde{y}_{\nu_i} = \frac{\tilde{c}}{m}\left\langle \frac{(M^{-1}x)^{\nu_i}}{\nu_i!}, \sum_{\beta\in\mathbb{Z}^d} A_\beta\Phi(x-\beta) \right\rangle.
$$

Therefore, Equation (3.24) yields

$$
\begin{aligned}
\tilde{y}_{\nu_i} &= \frac{\tilde{c}}{m}\left\langle \sum_{j=1}^N w(\nu_i,\nu_j)\frac{x^{\nu_j}}{\nu_j!}, \sum_{\beta\in\mathbb{Z}^d} A_\beta\Phi(x-\beta) \right\rangle \\
&= \frac{\tilde{c}}{m}\sum_{j=1}^N w(\nu_i,\nu_j)\sum_{\beta\in\mathbb{Z}^d} A_\beta\left\langle \frac{(x+\beta)^{\nu_j}}{\nu_j!}, \Phi(x) \right\rangle \\
&= \frac{\tilde{c}}{m}\sum_{j=1}^N w(\nu_i,\nu_j)\sum_{\beta\in\mathbb{Z}^d} A_\beta\sum_{\kappa\leq\nu_j} \frac{\beta^\kappa}{\kappa!}\left\langle \frac{x^{\nu_j-\kappa}}{(\nu_j-\kappa)!}, \Phi(x) \right\rangle.
\end{aligned}
$$

Consequently, we get

$$\widetilde{y}_{\nu_i} = \sum_{j=1}^{N} w(\nu_i, \nu_j) \sum_{\kappa \leq \nu_j} J_A(\kappa) \widetilde{y}_{\nu_j - \kappa}$$

which is equivalent to

$$\sum_{j=1}^{N} \big(\delta_{i,j} \mathbf{I}_r - w(\nu_i, \nu_j) J_A(0)\big) \widetilde{y}_{\nu_j} = \sum_{j=1}^{N} w(\nu_i, \nu_j) \widetilde{v}_A(\nu_j).$$

Thus, we have

$$\mathbf{P}_N \begin{pmatrix} \widetilde{y}_{\nu_1} \\ \vdots \\ \widetilde{y}_{\nu_N} \end{pmatrix} = \mathbf{Q}_N \begin{pmatrix} \widetilde{v}_A(\nu_1) \\ \vdots \\ \widetilde{v}_A(\nu_N) \end{pmatrix}.$$

It remains to be shown that $\mathbf{P}_N$ is invertible or, equivalently, that 1 is not an eigenvalue of $\mathbf{W}_N \otimes J_A(0)$. Define $g_N : \mathbb{R}^d \to \mathbb{R}^N$ via

$$g_N(x) := \left(\frac{x^{\nu_1}}{\nu_1!}, \ldots, \frac{x^{\nu_N}}{\nu_N!} \right)^{\mathsf{T}},$$

then Equation (3.24) implies that for $x \in \mathbb{R}^d$ and $n > 0$

$$\mathbf{W}_N^n g_N(x) = g_N(M^{-n}x).$$

Since M is expanding, M^{-1} has spectral radius $\mathrm{spr}(M^{-1}) < 1$. For a fixed θ with $\mathrm{spr}(M^{-1}) < \theta < 1$, standard results imply that there exists a constant C, depending on the norm $\|.\|$ only, such that

$$\|M^{-n}x\| \leq C\|x\|\theta^n, \quad x \in \mathbb{R}^d,\ n \geq 0,$$

cf. Lemma 1.3.3 in [104]. Thus, for large n and $x \in \mathbb{R}^d$ we obtain $|(M^{-n}x)_i| < 1$, $0 \leq i < d$, and therefore

$$\|\mathbf{W}_N^n g_N(x)\|_\infty = \max_{1 \leq j \leq N} \left| \frac{(M^{-n}x)^{\nu_j}}{\nu_j!} \right| \leq \|M^{-n}x\|_\infty.$$

As a consequence,

$$\mathbf{W}_N^n g_N(x) \longrightarrow 0 \quad \text{as } n \to 0. \tag{6.5}$$

It was shown in [106] that for each N–dimensional space of polynomials there exist nodes $x_1, \ldots, x_N$ such that the corresponding Lagrange interpolation problem is uniquely solvable. Therefore, since the monomials $x^{\nu_j}/\nu_j!$, $1 \leq j \leq N$, form a basis of the space of homogeneous polynomials of total degree n, there exist points $x_1, \ldots, x_N \in \mathbb{R}^d$ such that the vectors $g_N(x_j)$, $1 \leq j \leq N$, i.e., the rows of the corresponding Vandermonde matrix, are linearly independent. Hence, the set

$$\{g_N(x_j)|\, 1 \leq j \leq N\} \cup \{ig_N(x_j)|\, 1 \leq j \leq N\}$$

is a basis of $\mathbb{C}^N$. Expanding the eigenvectors of $\mathbf{W}_N$ with respect to this basis and applying (6.5) implies $\mathrm{spr}(\mathbf{W}_N) < 1$.

On the other hand, we have seen in Theorem 3.2.1 that the biorthogonality of Φ and $\tilde{\Phi}$ implies $\mathrm{spr}(\mathbf{A}(1)) = m$ and thus $\mathrm{spr}(J_A(0)) = 1$. A direct computation shows that the eigenvalues of $\mathbf{W}_N \otimes J_A(0)$ are pairwise products of the eigenvalues of $\mathbf{W}_N$ and the eigenvalues of $J_A(0)$. As a consequence, 1 cannot be an eigenvalue of $\mathbf{W}_N \otimes J_A(0)$. Thus, $\mathbf{P}_N$ is invertible. □

Once we are given $\tilde{y}_0$, Theorem 6.1.9 enables us to compute the vectors $\tilde{y}_\mu$ recursively. Since $\tilde{y}_0 = \hat{\tilde{\Phi}}(0)$, an eigenvector of $\mathbf{A}(1)$ corresponding to the eigenvalue m determines $c \cdot \tilde{y}_0$ for some constant c. This constant appears as a factor on both sides of the sum rules (3.23), thus we can incorporate the sum rules into our construction.

6.2 Explicit Construction

In this section, we give an explicit construction method for the masks of symmetric interpolating scaling vectors on $\mathbb{R}^d$ with compact support. In addition, we state an algorithm for constructing the masks of dual scaling vectors which are symmetric and compactly supported as well. In contrast to our construction of orthonormal interpolating scaling vectors in Section 5.2, there are only linear conditions involved in the following construction process. Therefore, in many cases, we are enabled to compute both, the primal and some corresponding dual masks analytically. To substantiate our approach, several examples for the case $d = 2$ are presented.

6.2.1 General Method

One of the main benefits of the biorthogonal approach is that the construction process is somewhat decoupled such that the primal and the dual scaling vectors or their masks, respectively, can be constructed consecutively. As we have seen in the preceding section, the primal and dual masks are connected by the biorthogonality condition in Proposition 6.1.8. Furthermore, the sum rules for the dual mask are determined by the corresponding primal mask by means of Theorem 6.1.9. Consequently, it suggest itself to construct the masks of the primal scaling vectors first.

Similar to the orthonormal case, the first step consists of choosing the basic parameters, i.e., the scaling matrix M and a complete set of representatives $R = \{0, \rho_1, \ldots, \rho_{m-1}\}$ of $\mathbb{Z}^d / M\mathbb{Z}^d$. Furthermore, an appropriate symmetry group $\mathcal{G}$ has to be chosen.

Primal Scaling Vectors

Based on the results in the preceding section, we suggest the following construction principle for symmetric interpolating scaling vectors:

1. To determine the support of the mask $A \in \ell_0(\mathbb{Z}^d)^{m \times m}$, choose the sets Ω_j in Proposition 6.1.7 for $1 \leq j < m$ and compute some minimal generating sets $\Lambda_j \subset \Omega_j$. Thus, we start with $m \cdot \sum_{j=1}^{m-1} |\Lambda_j|$ degrees of freedom.

2. Choose a proper sum rule order k (i.e. as high as possible) and solve the system of linear equations given by the sum rules in Theorem 5.1.6 with respect to the symmetry conditions in Corollary 6.1.3.

3. Find the best solution.

If the sets Λ_j are not too large, we have to deal with a moderate number of linear equations only. Hence, the system in step 2 can be solved analytically using a symbolic computation tool like Maple or MuPAD. In general, this system is underdetermined and thus we obtain a solution A_ϑ which depends on a parameter vector $\vartheta \in \mathbb{R}^t$ for some $t > 0$. Therefore, as step 3 of our scheme, we can use these remaining degrees of freedom to maximize the regularity of the corresponding scaling vector Φ_ϑ, i.e., we have to maximize the function

$$F : \mathbb{R}^t \ni \vartheta \longmapsto \mathfrak{s}(\Phi_\vartheta).$$

As we have seen in Section 3.2.3, computing the critical Sobolev exponent of a scaling vector is a rather delicate task. Hence, it is pretty unlikely to obtain any derivative information of the function F. As a consequence, for solving this unconstrained optimization problem we have to employ methods that are based on point evaluations only.

One appropriate method is the *downhill simplex method* introduced by Nelder and Mead in [95]. It is initialized by $t + 1$ noncollinear points in $\mathbb{R}^t$ which are taken to be the vertices of a nondegenerate t-dimensional simplex. Essentially, the downhill simplex method consists of iteratively moving those vertices of the simplex which belong to the smallest values of F. If after some iteration steps the function values of all vertices are close to each other, the algorithm stops.

Another method for solving the above optimization problem is the direction set method stated in [100], see also [4]. Similar to the downhill simplex method, it is initialized with $t+1$ vectors $\vartheta_0, \tau_1, \ldots, \tau_t \in \mathbb{R}^t$, but here the vectors $\tau_1, \ldots, \tau_t$ are direction vectors which should be orthogonal or at least linearly independent. Then, starting from the point $\vartheta_0^0 := \vartheta_0$ one consecutively solves the problems

$$F(\vartheta_0^i + s\tau_{i+1}) = \max, \quad 0 \leq i \leq t, \tag{6.6}$$

where in each step the point ϑ_0^i corresponds to the maximum of the preceding step. Afterwards, one sets $\vartheta_0 := \vartheta_0^t$, updates the directions $\tau_1, \ldots, \tau_t$ in a certain manner, and restarts the process. If, after a complete cycle of line maximizations (6.6), the gain $F(\vartheta_0^t) - F(\vartheta_0^0)$ drops below a preassigned tolerance, the algorithm terminates.

Remark 6.2.1. *Our construction scheme can also be used to construct scalar generators. If a mask satisfies*

$$a_\beta^{(i,j)} = a_{\beta-\rho_i}^{(0,j)}, \quad 0 \leq i,j < m, \tag{6.7}$$

then a direct computation shows that the corresponding scaling vector has to stem from a scaling function via the embedding (3.7). Thus, incorporating condition (6.7) into step 2 of our construction leads to scalar generators.

Dual Scaling Vectors

Given the mask of a symmetric interpolating scaling vector, the mask of a dual scaling vector can be obtained as follows:

1. For $0 \leq i < m$ choose the symmetry center c_i of $\widetilde{\phi}_i$. Due to the biorthogonality of Φ and $\widetilde{\Phi}$, the choice $c_i = M^{-1}\rho_i$ suggests itself.

2. Determine the support of $\widetilde{A} \in \ell_0(\mathbb{Z}^d)^{m \times m}$ by choosing $\widetilde{\Omega}_j$, $0 \leq j < m$, and compute some minimal generating sets $\widetilde{\Lambda}_j \subset \widetilde{\Omega}_j$ corresponding to Proposition 6.1.7. Thus, we have $m \cdot \sum_{j=0}^{m-1} |\widetilde{\Lambda}_j|$ degrees of freedom.

3. Apply the biorthogonality condition (6.2) to the coefficient sequence $(\widetilde{A}_\beta)_{\beta \in \mathbb{Z}^d}$ with respect to the symmetry conditions in Theorem 6.1.1.

4. Choose a proper sum rule order $\widetilde{k}$ and compute the vectors $\widetilde{y}_\mu$, $|\mu| < \widetilde{k}$, by Theorem 6.1.9.

5. Apply the sum rules of order $\widetilde{k}$ to the coefficient sequence $\widetilde{A}$ with respect to the symmetry conditions in Theorem 6.1.1.

6. Find the best solution.

Again, for sets $\widetilde{\Lambda}_j$ of moderate size we have to deal with a system of linear equations that can be solved analytically. In most cases, also this system is underdetermined and hence the remaining degrees of freedom can be used to maximize the regularity of the corresponding dual scaling vector by utilizing one of the methods stated above. If the primal scaling vector stems from a scalar function, incorporating condition (6.7) for the dual mask into steps 3–5 allows the construction of scalar duals.

Since the number of vanishing moments of the primal multiwavelets is determined by the accuracy of the dual scaling vector, cf. Section 3.2.2, the sum rule order $\widetilde{k}$ in step 4 should be at least as large as k.

Numerical Issues

In order to apply the above construction schemes proficiently, some comments about numerical or technical details are required. First of all, we have to decide which of the two optimization methods should be prefered. Although it is commonly known that the direction set method almost surely converges faster than the downhill simplex method, in our setting we do not always share this observation. For a small number of parameters ($t \leq 3$), the direction set method indeed seems to converge faster. However, for larger t, the downhill simplex method almost always performs better than the direction set method. We assume that this observation is due to the fact that the initialization of the downhill simplex method is somewhat more amenable. Since we can roughly estimate the order of magnitude of the entries of θ, we can choose reasonable vertices for the starting simplex. Although we can also choose a reasonable starting point for the direction set method, a proper choice of directions is somewhat more involved. In spite of these observations, it is commonly known that for high-dimensional problems both algorithms converge rather slowly.

On the other hand, both methods converge locally, i.e., they lead to local maxima only. Therefore, for a given problem, we suggest to apply either of these methods repeatedly with distinct initialization parameters. Moreover, especially for large t, it may pay to use both methods alternatingly with mutually adapted parameters. Nevertheless, in contrast to our construction of orthonormal scaling vectors in Section 5.2, each fixed $\vartheta \in \mathbb{R}^t$ corresponds to an exact solution of our system of equations and therefore leads to a scaling vector Φ_ϑ. Consequently, being stuck in a local maximum is not too serious. Apart from suboptimal regularity results, this does not spoil the applicability of the corresponding scaling vector.

Another problem we have to address is the influence of the primal solution on the outcome of the dual construction process. As we have seen above, most of the conditions involved in the construction of dual scaling vectors are determined by the mask of the primal scaling vector. Since the parameter vector ϑ of the primal solution is obtained numerically, its entries usually are floating point numbers of which a symbolic representation is not known. Hence, trying to solve the dual system of equations with a symbolic computation tool is likely to fail. Therefore, to facilitate the dual construction process, we suggest to postprocess the primal solution. In particular, we observe that approximating the parameters ϑ by rational numbers such that the divisor is a (small) power of 2 leads to good results. Of course, one has to assure that the regularity of the primal scaling vector does not deteriorate too much.

Sufficient Conditions

Similar to our approach in the preceding chapter, the conditions on the mask involved in our construction method are necessary only. Therefore, during the construction process, one again has to check whether the corresponding scaling vectors actually do possess the desired properties. This can be performed analogously to the orthonormal setting, cf. Section 5.2.

To ensure biorthogonality, one first has to check that $\mathbf{A}(1)$ and $\widetilde{\mathbf{A}}(1)$ satisfy condition (iii) of Theorem 3.2.1. As in the orthormal setting, condition (iv) and $\Phi, \widetilde{\Phi} \in L_2(\mathbb{R}^d)^m$ can be verified as a byproduct of the regularity estimation.

Checking the interpolation property of the primal scaling vector has already been discussed in Section 5.2.

6.2.2 Examples

For all our examples, we observe that introducing the additional condition

$$\sum_{j=0}^{m-1} \sum_{\beta \in \mathbb{Z}^d} a_\beta^{(i,j)} = m \tag{6.8}$$

or, equivalently, $\widetilde{y}_0 = (1, \ldots, 1)^\top$ in step 2 of the primal construction process considerably improves the outcome of the dual process. The reason for this may be twofold. First of all, we have seen in Section 6.1.3 that, once given $\widetilde{y}_0$, the vectors $\widetilde{y}_\mu$ for the dual sum rules can be computed recursively. Therefore, similar to the postprocessing of the primal solutions stated above, a *nice* vector $\widetilde{y}_0$ may lead to more amenable conditions. On the other hand, $\widetilde{y}_0 = (1, \ldots, 1)^\top$ implies that the primal scaling vector is *balanced* of order 1, see Section 7.1.3 for details. This might have a positive influence on the properties of the dual solution. Apart from that, we observe that the regularity of scaling vector corresponding to the primal solution becomes only marginally lower by introducing the additional condition (6.8).

Moreover, in all our examples it turns out that if the choice of Λ_1 and k yields the mask of a primal scaling vector, then also a mask satisfying condition (6.7) can be obtained. Hence, we obtain vector valued primal solution and, in addition, scalar primal solutions which possess similar support properties and provide the same approximation order. For simplicity of notation, we consider the scalar solutions as scaling vectors via the embedding (3.7). The most relevant masks of the obtained examples can be found in Appendix A.3.

Example 1

In our first example, we construct interpolating scaling vectors in $L_2(\mathbb{R}^2)^2$ for the well–known quincunx matrix

$$M_q = \begin{pmatrix} 1 & -1 \\ 1 & 1 \end{pmatrix}$$

which has already been introduced in Section 5.2.2. We choose the nontrivial representative $\rho_1 = (0,1)^\top$, and it can be shown that

$$\mathcal{G} := \left\{ \pm \mathbf{I}, \pm \begin{pmatrix} 0 & -1 \\ 1 & 0 \end{pmatrix}, \pm \begin{pmatrix} 1 & 0 \\ 0 & -1 \end{pmatrix} \right\}$$

forms a proper symmetry group. Since the set Ω_1 has to be $\mathcal{G}$–symmetric with respect to $-M^{-1}\rho_1 = -(1/2, 1/2)^\top$, we choose $\Omega_1 := [-n, n-1]^2 \cap \mathbb{Z}^2$ and $\Lambda_1 := \{(\beta_0, \beta_1)^\top \in \Omega_1 \mid 0 \le \beta_1 \le \beta_0\}$ for some $n \in \mathbb{N}$.

n	sum rule	$\mathfrak{s}(\Phi)$		
	order	Φ_n^{sc}	Φ_n^v	Φ_n^∞
1	2	1.578		1.662
2	4	2.628	2.637	2.748
3	6	3.338	3.664	3.714
4	8	4.238	4.553	4.735

Table 6.1: Properties of the Φ_n for M_q

Table 6.1 shows the properties of the outcome of our construction for $n = 1, \ldots, 4$. We use the notation Φ_n^∞ for the best solutions and Φ_n^v for the vector valued solutions satisfying (6.8). The embedded scalar solutions are denoted by Φ_n^{sc}. For $n = 1$ the solutions Φ_1^{sc} and Φ_1^v coincide, they correspond to the well known Laplace symbol, see [24] for details. In addition to Φ_2^{sc} we obtain a scalar solution Φ_2^{sc*} with Sobolev regularity $s = 2.50$ which has already been constructed in [30]. Table 6.1 shows that for a fixed n the vector valued solutions are slightly smoother than the scalar solutions. The gain is about 10%.

ϕ_0

ϕ_1

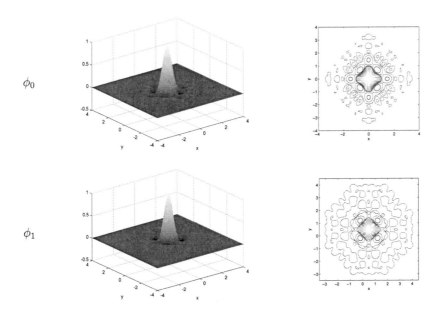

Figure 6.1: Component functions of Φ_3^v for M_q

ϕ_0

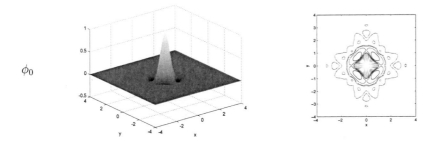

Figure 6.2: First component function of Φ_3^{sc} for M_q

In Figure 6.1 the component functions of Φ_3^v are shown. For comparison, also the first component function of the scalar solution Φ_3^{sc}, taken as a scaling vector via the embedding (3.7), is shown in Figure 6.2. All these functions possess the desired symmetry properties and are very similar in shape. However, the components of the true scaling vector seem to provide a stronger oscillation behaviour.

For the dual functions, the sets $\widetilde{\Omega}_0$ and $\widetilde{\Omega}_1$ have to be $\mathcal{G}$–symmetric with centers 0 and $-M^{-1}\rho_1$, respectively. Therefore, we choose $\widetilde{\Omega}_0 := [-\widetilde{m}, \widetilde{m}]^2 \cap \mathbb{Z}^2$ and $\widetilde{\Lambda}_0 := \{(\beta_0, \beta_1)^\top \in \widetilde{\Omega}_0 \,|\, 0 \le \beta_1 \le \beta_0\}$ for some $\widetilde{m} \in \mathbb{N}$. Furthermore, we define $\widetilde{\Omega}_1 := \Omega_1$ and $\widetilde{\Lambda}_1 := \Lambda_1$ for some $\widetilde{n} \in \mathbb{N}$. In the following examples we focus on the case $\widetilde{k} \le k$ only.

primal	$\widetilde{m}$	$\widetilde{n}$	sum rule order	$\mathfrak{s}(\widetilde{\Phi})$	
				$\widetilde{\Phi}_{\widetilde{m},\widetilde{n}}^{sc}$	$\widetilde{\Phi}_{\widetilde{m},\widetilde{n}}^{v}$
Φ_1^{sc}	2	2	2	0.749	1.110
Φ_2^{sc}	3	2	4	0.852	1.152
Φ_3^{sc}	4	2	4	0.941	1.289
Φ_3^{sc}	5	3	6	1.394	1.959
Φ_4^{sc}	5	2	4	0.969	1.330
Φ_4^{sc}	6	3	6	1.495	2.276
Φ_4^{sc}	7	4	8	2.130	2.953

Table 6.2: Properties of the dual functions for Φ^{sc} and M_q

Table 6.2 shows the properties of the solutions corresponding to our scalar primal generators. Again, we obtain scalar as well as vector valued dual functions denoted by $\widetilde{\Phi}_{\widetilde{m},\widetilde{n}}^{sc}$ and $\widetilde{\Phi}_{\widetilde{m},\widetilde{n}}^{v}$, respectively. In contrast to the primal case, the vector setting yields a gain of about 40% in regularity. Moreover, all the obtained vector valued duals are continuous.

In the literature, several examples of dual functions related to our scalar primal solutions can be found, see, e.g., [32, 68]. To compare these functions with our results, they have to be translated into our setting. First of all, by means of the rule (3.7), these scalar generators can be considered as scaling vectors $\widetilde{\Phi}^{sc}$. Furthermore, all these functions are symmtric about the origin. Therefore, due to Theorem 6.1.1, each mask is symmetric with the origin as the symmetry center. As a consequence, the support of such a mask is centered about its zeroth coefficient which perfectly fits into our setting. Thus, all these duals can be considered as scaling vectors $\widetilde{\Phi}_{\widetilde{m},\widetilde{n}}^{sc}$ with support parameters $\widetilde{m}$ and $\widetilde{n}$ as above.

In [68] a collection of duals for the primal generator Φ_1^{sc} have been constructed. There, the most regular dual of Φ_1^{sc} is continuous but not differentiable and, translated into our setting, corresponds to a scaling vector $\widetilde{\Phi}_{\widetilde{m},\widetilde{n}}^{sc}$ with support parameters

$\widetilde{m} = \widetilde{n} = 4$. In contrast, our approach leads to a dual scaling vector $\widetilde{\Phi}^v_{2,2}$ which is continuous as well. Another interesting example can be found in [32]. There, a continuous dual of Φ^{sc*}_2 has been constructed. Translated to our setting, it corresponds to a scaling vector $\widetilde{\Phi}^{sc}_{\widetilde{m},\widetilde{n}}$ with support parameters $\widetilde{m} = \widetilde{n} = 6$. Again, our approach leads to a true vector valued dual $\widetilde{\Phi}^v_{\widetilde{m},\widetilde{n}}$ with similar regularity properties but smaller support parameters $\widetilde{m} = 3$ and $\widetilde{n} = 2$. However, it should be mentioned that the dual scaling functions in [68] and [32] provide a higher approximation order than our solutions.

primal	$\widetilde{m}$	$\widetilde{n}$	sum rule order	$\mathfrak{s}\big(\widetilde{\Phi}^v_{\widetilde{m},\widetilde{n}}\big)$
Φ^∞_1	2	2	2	1.265
Φ^v_2	3	2	4	1.196
Φ^v_3	4	2	4	1.316
Φ^v_3	5	3	6	2.391
Φ^v_4	5	2	4	1.676
Φ^v_4	6	3	6	2.335
Φ^v_4	7	4	8	2.548

Table 6.3: Properties of the dual functions for Φ^v and M_q

In Table 6.3 the properties of some duals corresponding to the primal scaling vectors are stated. As already mentioned above, for the Φ^∞_n we obtain duals that are less regular then those of the Φ^{sc}_n, e.g., $\widetilde{\Phi}^v_{5,3} \in H^{1.85}$ corresponding to Φ^∞_3. On the other hand, the Φ^v_n lead to dual scaling vectors which are very smooth, even smoother than the vector valued duals of the scalar Φ^{sc}_n. For example, for Φ^v_3 we obtain a dual scaling vector $\widetilde{\Phi}^v_{5,3}$ which is continuously differentiable. Unfortunately, we observe that the dual $\widetilde{\Phi}^v_{7,4}$ of Φ^v_4 is less regular than the dual $\widetilde{\Phi}^{sc}_{7,4}$ of Φ^{sc}_4, and therefore does not meet our expectations. However, since for all other examples the duals of the vector valued primals Φ^v_n are considerably smoother than the duals of the scalar primals Φ^{sc}_n, the reason for this misbehaviour may be of numerical nature. Indeed, during the construction of the dual $\widetilde{\Phi}^v_{7,4}$ of Φ^v_4, we are confronted with several numerical problems. First of all, the solution of the corresponding equation system reveals a large number of parameters, i.e., we obtain a mask $\widetilde{A}_\vartheta$ depending on a parameter vector $\vartheta \in \mathbb{R}^t$ with $t = 11$. Therefore, as already stated above, the downhill simplex method as well as the direction set method converge, if at all, rather slowly. Furthermore, we observe that both algorithms frequently run into local maxima such that we have to restart the process quite often. On the other hand, due to the support size of the mask, computing the critical Sobolev exponent of the corresponding scaling vector is very time consuming, cf. Section

3.2.3. As a consequence, employing several computers (including also the cluster of the Marburg University Computing Center) for many days led us to the example stated above. However, it is worth mentioning that we found several examples with similar regularity properties for distinct ϑ. Hence, we assume that using more sophisticated optimization methods this problem can be bypassed.

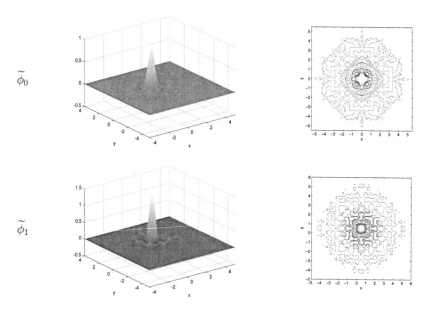

Figure 6.3: Component functions of $\widetilde{\Phi}^v_{5,3}$ for M_q

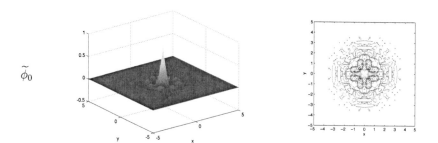

Figure 6.4: First component function of $\widetilde{\Phi}^{sc}_{5,3}$ for M_q

Examples of some dual functions are depicted in Figure 6.3 and in Figure 6.4. In Figure 6.3 the component functions of the dual scaling vector $\widetilde{\Phi}^v_{5,3}$ of Φ^v_3 are shown. Figure 6.4 shows a component function of $\widetilde{\Phi}^{sc}_{5,3}$, a scalar dual of Φ^{sc}_3, taken as a scaling vector via (3.7). Similar to the primal generators, all functions show the desired symmetry properties, and the vector valued function is visibly smoother than its scalar counterpart.

Example 2

For our second example we choose the scaling matrix

$$M_s := \begin{pmatrix} 0 & -1 \\ 2 & 0 \end{pmatrix}$$

which generates the lattice $M_s \mathbb{Z}^2 = \mathbb{Z} \times 2\mathbb{Z}$. Thus, we can choose the nontrivial representative $\rho_1 := (0,1)^\top$. Obviously, the set

$$\mathcal{G} := \left\{ \pm \mathbf{I}, \pm \begin{pmatrix} 1 & 0 \\ 0 & -1 \end{pmatrix} \right\}$$

forms a proper symmetry group. Furthermore, we choose $\Omega_1 := ([-n, n-1] \times [-n, n]) \cap \mathbb{Z}^2$ and $\Lambda_1 := \{\beta \in \Omega_1 \,|\, \beta \geq 0\}$ for some $n \in \mathbb{N}$. Similar to our first example, the most regular scaling vectors Φ^∞_n lead to poor dual functions. Therefore, we focus on scalar solutions Φ^{sc}_n and vector valued solutions Φ^v_n which satisfy (6.8). Table 6.4 shows that the primal solutions for the scaling matrix M_s

n	sum rule order	$\mathfrak{s}(\Phi \in H^s)$	
		Φ^{sc}_n	Φ^v_n
1	2	1.575	
2	4	2.569	2.823
3	6	3.300	3.660

Table 6.4: Properties of the Φ_n for M_s

and for the quincunx matrix M_q possess very similar properties. For $n = 1$, the solutions Φ^{sc}_1 and Φ^v_1 coincide and for $n > 1$, the Φ^v_n are slightly smoother than the Φ^{sc}_n.

In Figure 6.5 the component functions of a typical example of the primal scaling vectors corresponding to M_s are depicted. Both functions possess the desired symmetry properties and show rectangular structures which correspond to the stripes grid induced by M_s.

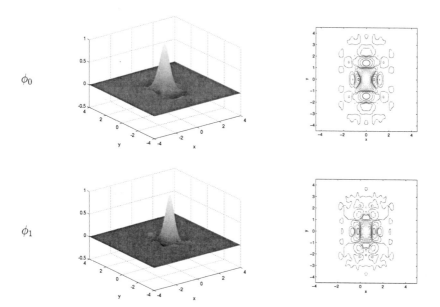

Figure 6.5: Component functions of Φ_3^v for M_s

primal	m	n	sum rule order	$\mathfrak{s}(\widetilde{\Phi} \in H^s)$	
				$\widetilde{\Phi}_{m,n}^{sc}$	$\widetilde{\Phi}_{m,n}^{v}$
Φ_1^{sc}	2	2	2	1.150	1.150
Φ_2^{sc}	3	2	4	0.797	0.943
Φ_2^{v}	3	2	4	—	1.044
Φ_3^{sc}	5	3	6	1.576	2.098
Φ_3^{v}	5	3	6	—	2.142

Table 6.5: Properties of the dual functions for M_s

For the dual functions we choose $\widetilde{\Omega}_0 := [-\widetilde{m}, \widetilde{m}]^2 \cap \mathbb{Z}^2$ and $\widetilde{\Lambda}_0 := \{(\beta \in \widetilde{\Omega}_0 \,|\, \beta \geq 0\}$. Similar to Example 1, we define $\widetilde{\Omega}_1 := \Omega_1$ and $\widetilde{\Lambda}_1 := \Lambda_1$ for some $\widetilde{n} \in \mathbb{N}$, and focus on the case $\widetilde{k} = k$ only.

The properties of the dual functions are shown in Table 6.5. For $n = 1$, both duals of Φ_1^{sc} possess the same regularity though $\widetilde{\Phi}_{2,2}^{sc}$ and $\widetilde{\Phi}_{2,2}^{v}$ are not identical. However, for larger n the results are similar to those of the quincunx case, i.e., the vector valued duals considerably outperform their scalar counterparts.

6.3 Multiwavelets

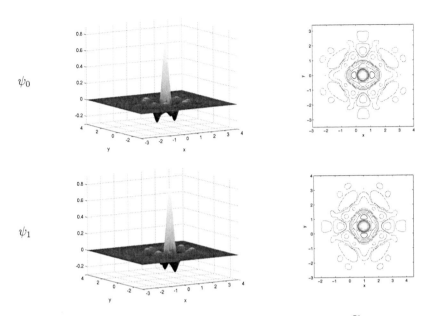

ψ_0

ψ_1

Figure 6.6: Primal multiwavelets corresponding to $(\Phi_3^v, \widetilde{\Phi}_{5,3}^v)$ for M_q

Let $(\Phi, \widetilde{\Phi})$ be a pair of compactly supported biorthogonal r–scaling vectors with masks $A, \widetilde{A} \in \ell_0(\mathbb{Z}^d)^{r \times r}$. We have seen in Theorem 3.2.2 that the biorthogonal multiwavelets $\Psi^{(n)}$ and $\widetilde{\Psi}^{(n)}$, $1 \leq n < m$, can be obtained by solving a matrix extension problem related to the modulation matrices $\mathcal{P}_m(z)$ and $\widetilde{\mathcal{P}}_m(z)$ defined in Equation (3.19). With the subsymbol notation introduced in (3.4) we define the *polyphase matrix* of Φ by

$$\mathcal{P}(z) := \begin{pmatrix} \mathbf{A}_0(z) & \cdots & \mathbf{A}_{m-1}(z) \\ \mathbf{B}_0^{(1)}(z) & \cdots & \mathbf{B}_{m-1}^{(1)}(z) \\ \vdots & \ddots & \vdots \\ \mathbf{B}_0^{(m-1)}(z) & \cdots & \mathbf{B}_{m-1}^{(m-1)}(z) \end{pmatrix}.$$

A direct computation using Equations (3.4) and (3.5) shows that the extension problem in Theorem 3.2.2 can equivalently be stated in terms of the polyphase matrices, i.e., Equation (3.20) holds if and only if

$$\mathcal{P}(z)\overline{\widetilde{\mathcal{P}}(z)}^{\top} = m\mathbf{I}_{rm}, \tag{6.9}$$

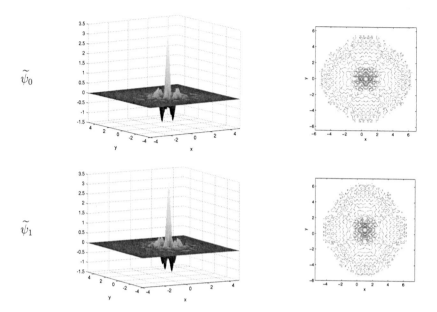

Figure 6.7: Dual multiwavelets corresponding to $(\Phi_3^v, \widetilde{\Phi}_{5,3}^v)$ for M_q

where $\widetilde{\mathcal{P}}(z)$ denotes the dual polyphase matrix.

In the scalar setting, this problem can be solved with the extension algorithm stated in [102]. The first step of this algorithm utilizes the well-known Quillen–Suslin Theorem to find an invertible extension of the first row of $\mathcal{P}(z)$. Then, the rows of the extended matrix are orthogonalized to obtain $\mathcal{P}(z)$. In the vector case, however, not only a row vector but an $r \times mr$-matrix has to be extended, and thus it is much more complicated to apply the Quillen–Suslin Theorem. Hence, in the known literature there is no fundamental method for obtaining some multiwavelets for given multivariate scaling vectors. Fortunately, due to the simple structure (3.10) of the symbol of an interpolating scaling vector, in our case the invertible extension can be found by inspection. Then, the orthogonalization step can be performed by using block matrices. As a consequence, we obtain the following theorem which appears as one of the main results of this work.

Theorem 6.3.1. *For $1 \le i < m$ define $C_i := (C_{i,1}, \dots, C_{i,m})$ with $C_{i,j} := e_i^\top \otimes e_j$, where e_i denotes the ith unit vector in $\mathbb{R}^m$. Furthermore, let $\mathcal{P}_1(z)$ and $\widetilde{\mathcal{P}}_1(z)$ denote the first $m \times m^2$-blocks of $\mathcal{P}(z)$ and $\widetilde{\mathcal{P}}(z)$, respectively, where $\mathcal{P}_1(z)$ corresponds to a compactly supported interpolating m–scaling vector. If we define the* *ith $m \times m^2$-block of $\mathcal{P}(z)$, $1 < i \leq m$, by*

$$\mathcal{P}_i(z) := C_i - \frac{1}{m} C_i \overline{\widetilde{\mathcal{P}}_1(z)}^\top \mathcal{P}_1(z)$$

then $\mathcal{P}(z)$ is nonsingular on $\mathbb{C}^d \setminus \{0\}$ and $\mathcal{P}(z)\overline{\widetilde{\mathcal{P}}_1(z)}^\top = (\mathbf{I}_m, 0)^\top$.

Proof. A direct computation yields that $\mathcal{P}(z)\overline{\widetilde{\mathcal{P}}_1(z)}^\top = (\mathbf{I}_m, 0)^\top$. It remains to be shown that $\mathcal{P}(z)$ is nonsingular. Let $d_i := (\lambda_{i,1}, \ldots, \lambda_{i,m})^\top \in \mathbb{R}^m$, $1 \leq i \leq m$, with $\lambda_{i,j} \neq 0$ for some i, j and let $z \in \mathbb{C}^d \setminus \{0\}$ such that

$$0 = (d_1^\top, \ldots, d_m^\top)\mathcal{P}(z).$$

Thus, we have

$$
\begin{aligned}
0 &= \sum_{i=1}^m d_i^\top \mathcal{P}_i(z) \\
&= d_1^\top \mathcal{P}_1(z) + \sum_{i=2}^m d_i^\top \left(C_i - \frac{1}{m} C_i \overline{\widetilde{\mathcal{P}}_1(z)}^\top \mathcal{P}_1(z) \right) \\
&= \left(d_1^\top - \frac{1}{m} \sum_{i=2}^m d_i^\top C_i \overline{\widetilde{\mathcal{P}}_1(z)}^\top \right) \mathcal{P}_1(z) + \sum_{i=2}^m d_i^\top C_i. \qquad (6.10)
\end{aligned}
$$

Due to the interpolation condition (3.10), the first column of each $\mathbf{A}_j(z)$ is a unit vector e_j. Hence, the matrix

$$
\mathcal{Q}(z) := \begin{pmatrix} \mathbf{A}_0(z) & \cdots & \mathbf{A}_{m-1}(z) \\ C_{1,1} & \cdots & C_{1,m} \\ \vdots & \ddots & \vdots \\ C_{m,1} & \cdots & C_{m,m} \end{pmatrix}
$$

is nonsingular for all $z \in \mathbb{C}^d \setminus \{0\}$. Therefore, (6.10) is equivalent to

$$0 = \left(d_1^\top - \frac{1}{m} \sum_{i=2}^m d_i^\top C_i \overline{\widetilde{\mathcal{P}}_1(z)}^\top \right) \mathcal{P}_1(z)\mathcal{Q}(z)^{-1} + \sum_{i=2}^m d_i^\top C_i \mathcal{Q}(z)^{-1}.$$

Since $\mathcal{P}_1(z)\mathcal{Q}(z)^{-1} = (\mathbf{I}_m, 0)$ and $C_i \mathcal{Q}(z)^{-1} = e_i^\top \otimes \mathbf{I}_m$, we have

$$0 = \left(d_1^\top - \frac{1}{m} \sum_{i=2}^m d_i^\top C_i \overline{\widetilde{\mathcal{P}}_1(z)}^\top, 0, \ldots, 0 \right) + \left(0, d_2^\top, \ldots, d_m^\top \right).$$

Consequently, $\lambda_{i,j} = 0$ for $1 \leq i, j \leq m$, i.e., the rows of $\mathcal{P}(z)$ are linearly independent. $\qquad \square$

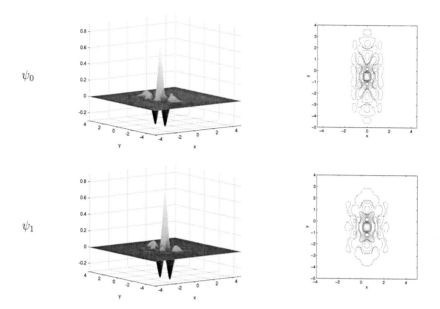

Figure 6.8: Primal multiwavelets corresponding to $(\Phi_3^v, \widetilde{\Phi}_{5,3}^v)$ for M_s

This theorem provides us with a painless way of computing some *canonical* primal multiwavelets. In addition, following the lines of the proof of Lemma 3.2.8 in [104], we obtain the following corollary. For the reader's convenience, we sketch the proof in our setting.

Corollary 6.3.2. *The polyphase matrix $\mathcal{P}(z)$ satisfies*

$$\det(\mathcal{P}(z)) = c \cdot z^\alpha$$

for some $\alpha \in \mathbb{Z}^d$ and $c \in \mathbb{R}$ with $c \neq 0$.

Proof. Due to Theorem 6.3.1 we have $\det(\mathcal{P}(z)) \neq 0$ on $\mathbb{C}^d \setminus \{0\}$. Since $\det(\mathcal{P}(z))$ is a Laurent polynomial, there exists a $\mu \in \mathbb{Z}_+^d$ such that

$$p(z) := z^\mu \det(\mathcal{P}(z))$$

is a polynomial on $\mathbb{C}^d$ with $p(z) \neq 0$ for $z \in \mathbb{C}^d \setminus \{0\}$ and $p(0) = 0$. Let I_p be the ideal generated by p, then the set of common zeros of all polynomials in I_p (or the *affine variety* of I_p) contains only $0 \in \mathbb{C}^d$. Hilbert's Nullstellensatz states that if q is some polynomial on $\mathbb{C}^d$ which vanishes on the variety of I_p, then there exists an $n \in \mathbb{Z}_+$ such that $q^n \in I_p$. Thus, with $q(z) := z^{\mathbf{1}}$, we have $z^{(n \cdot \mathbf{1})} \in I_p$ for some $n \in \mathbb{Z}_+$, i.e., there exists a polynomial h on $\mathbb{C}^d$ such that

$$z^{(n \cdot \mathbf{1})} = p(z)h(z).$$

On the other hand, each divisor of $z^{(n \cdot \mathbf{1})}$ is a monomial times a constant. Therefore, we have $p(z) = c \cdot z^\nu$ for some $\nu \in \mathbb{Z}_+^d$ and $c \in \mathbb{C} \setminus \{0\}$. As a consequence, we obtain

$$\det(\mathcal{P}(z)) = c \cdot z^{\nu - \mu} =: c \cdot z^\alpha.$$

Since the coefficients of the entries of $\mathcal{P}(z)$ are real-valued, we have $c \in \mathbb{R}$.   □

This corollary enables us to compute $\widetilde{\mathcal{P}}(z) = \mathcal{P}(z)^{-1}$ directly by applying Cramer's rule. Furthermore, it ensures that $\widetilde{\mathcal{P}}(z)$ consists of Laurent polynomials. This provides us with the dual multiwavelets.

Figure 6.6 and Figure 6.7 show the component functions of the primal and dual multiwavelets corresponding to the pair $(\Phi_3^v, \widetilde{\Phi}_{5,3}^v)$ for M_q. It turns out that all our quincunx multiwavelets are *mutually* symmetric. Though each component function itself shows some reflection symmetries, one component function seems to be a shifted and rotated version of the other. This observation has been been validated numerically. In contrast, the component functions of the multiwavelets corresponding to the scaling matrix M_s are less intertwined. For example, Figure 6.8 shows that each of the components of the primal multiwavelet corresponding to the pair $(\Phi_3^v, \widetilde{\Phi}_{5,3}^v)$ for M_s is symmetric on its own.

Chapter 7

Dessert: Application to Image Compression

In this chapter we study the suitability of our wavelets and multiwavelets for signal processing purposes. Since most of the examples of multiwavelets constructed in this work are bivariate, we focus on two-dimensional signals called *digital images*. As a test application we consider image compression. It has turned out in a series of papers [7, 8, 9, 113] that multiwavelet algorithms can cope with classical wavelet algorithms in image compression at least for the univariate case using a tensor product approach. Thus, we want to investigate the following questions:

(Q1) Are our univariate multiwavelets as well capable of providing a good compression performance?

(Q2) It is well known that the tensor product approach leads to the *preferred directions* phenomenon, i.e., image features along horizontal or vertical directions are reproduced better than features in other direction. So what does the nonseparable scaling matrix approach buy?

(Q3) In Chapter 6, we have constructed scalar as well as vector valued wavelets. Which class performs better in image analysis?

This chapter is organized as follows. First of all, we give a brief introduction to the theory of signal compression with a focus on transform based coding schemes. For a detailed discussion of this topic, see, e.g., the textbook [116], Chapter 11 in [88], and Chapter 10 in [112]. In addition, we recall the basic definition of the discrete multiwavelet transform which appears as a generalization of the classical discrete wavelet transform. Furthermore, we address some technical questions which concern the restriction of the transform to finite domains and the initialization problem. The latter is closely related to the notion of balancing. Finally, in the main part of this chapter, we compare the image compression performance of our wavelets and multiwavelets with that of several established wavelets, including one biorthogonal wavelet pair that is used within the JPEG2000 image compression standard [2].

7.1   Signal/Image Compression

In signal or image processing one is concerned with data consisting of a finite sequence $X \in \ell(\Lambda)$, $\Lambda \subset \mathbb{R}^d$ with $|\Lambda| < \infty$. Usually, one assumes that X is obtained from a function $f \in L_2(\mathbb{R}^d)$ by either sampling f or taking local averages of f. In the following, we focus on the sampling approach, i.e., we assume that f is at least continuous such that $X_\lambda = f(\lambda)$ for a finite section Λ of some lattice in $\mathbb{R}^d$. If f is smooth in $\lambda_0 \in \Lambda$, then the values of f in a small neighborhood of λ_0 are close to $f(\lambda_0)$. Thus, the signal X contains some correlation or redundancy whenever the set Λ is sufficiently dense. The aim of signal compression is to reduce this redundancy in order to be able to store or transmit signal data in an efficient form.

In general, a compression algorithm consists of an *encoder* $\mathcal{E} : \ell(\Lambda) \rightarrow \{0,1\}^*$ and a *decoder* $\mathcal{D} : \{0,1\}^* \rightarrow \ell(\Lambda)$, where $\{0,1\}^*$ denotes the set of all binary words, i.e., all finite sequences with values in $\{0,1\}$. The pair $(\mathcal{E}, \mathcal{D})$ is also called a *codec*. Its objective is that for all X belonging to a certain class of signals, $\Xi := \mathcal{E}(X)$ contains as few information as possible, i.e., $\Xi \in \{0,1\}^n$ for a small n. Furthermore, the reconstructed signal $\widetilde{X} := \mathcal{D}(\Xi)$ should reveal a low distortion rate. Two commonly used distortion or error measures are the *mean square error*

$$\mathrm{mse}(X, \widetilde{X}) := \frac{1}{|\Lambda|} \sum_{\lambda \in \Lambda} (X_\lambda - \widetilde{X}_\lambda)^2$$

and the *mean absolute difference*

$$\mathrm{mad}(X, \widetilde{X}) := \frac{1}{|\Lambda|} \sum_{\lambda \in \Lambda} |X_\lambda - \widetilde{X}_\lambda|.$$

Although the mse or other ℓ_2/L_2–related distortion measures find a widespread use in signal processing literature, it has been proposed by DeVore et al. in [45] to use ℓ_1/L_1–related measures like the mad in image compression, since they better reflect the visual quality of a reconstructed image. However, it has been pointed out in Chapter 10 of [112] that for high compression rates the above objective measures may not resemble subjective measures obtained by visual perception at all.

7.1.1 Transform Coding

A commonly known class of signal compression algorithms are *transform coding schemes* which consist of several steps as is shown in Figure 7.1. Within the first

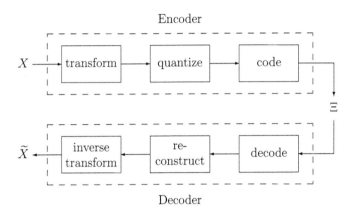

Figure 7.1: Transform coding scheme

step of the encoding scheme, the signal X is decorrelated, i.e., it is transformed into a signal consisting of few large coefficients which contain much information and many small coefficients with a low information content. In general, this step essentially reshuffles the information distribution among the coefficients but does not alter the number of coefficients. Hence, the amount of data is not reduced within this step. For example in image compression, where one often has to deal with discrete valued or indexed data like 8-bit grayscale images, this step does even increase the amount of data since the outcome of the transform may have arbitrary values. Therefore, in a quantization step, the range of values of the transformed signal is discretized and turned into a binary word or a bit stream. This can be performed by dividing the range of values in several classes and mapping each class onto a corresponding binary representation. Another possibility of quantization is to sort the values of the transformed signal, e.g., by magnitude or spatial relation, and then successively generate a bit stream by extracting the most significant bits of the values within this list. In this context, most significant bits means those bits in the binary representation of a value which belong to the largest powers of two. A very efficient version of this quantization method has been proposed in [97], see also [110]. Finally, the raw binary outcome of the quantization step can be encoded using one of the classical compression strategies like entropy coding and run-length encoding, see again [116] for details.

Within the decoding scheme, the operations performed in the encoding process have to be inverted. In general, the coding step as well as the transform step are invertible. However, this may not be the case for the quantization step, and thus one only obtains an approximation $\tilde{X}$ of the original signal X. If the quantization step is invertible, the codec is called *lossless*, otherwise it is called *lossy*.

The performance of a transform coding scheme is vastly influenced by the transform used to decorrelate the signal. In practice, one often uses the discrete Fourier transform or the discrete cosine transform, e.g., as in the JPEG standard [1]. In recent years, also the discrete wavelet transform has found its way into compression algorithms within the JPEG2000 standard [2].

7.1.2   Discrete Multiwavelet Transform

Assume we are given a biorthogonal pair of r–scaling vectors $(\Phi, \tilde{\Phi})$ with masks $A, \tilde{A} \in \ell_0(\mathbb{Z}^d)^{r \times r}$ Furthermore, assume we have a corresponding pair of biorthogonal r–multiwavelets $(\Psi^{(n)}, \tilde{\Psi}^{(n)})$, $1 \le n < m$, with masks $B^{(n)}, \tilde{B}^{(n)} \in \ell_0(\mathbb{Z}^d)^{r \times r}$. For a sequence $c_J := (c_{J,\beta})_\beta \in \ell_0(\mathbb{Z}^d)^r$, $J \in \mathbb{Z}$, the *discrete multiwavelet transform* (DMWT) is defined by the *analysis equations*

$$
\begin{aligned}
c_{j-1,\alpha} &:= \frac{1}{\sqrt{m}} \sum_\beta \tilde{A}_{\beta - M\alpha} c_{j,\beta} \\
d_{j-1,\alpha,n} &:= \frac{1}{\sqrt{m}} \sum_\beta \tilde{B}^{(n)}_{\beta - M\alpha} c_{j,\beta}
\end{aligned}
\tag{7.1}
$$

for $\alpha \in \mathbb{Z}^d$, $j = J, J-1, \ldots$, and $1 \le n < m$. The inverse DMWT is given by the *synthesis equation*

$$
c_{j,\alpha} = \frac{1}{\sqrt{m}} \sum_\beta \left(A^\top_{\alpha - M\beta} c_{j-1,\beta} + \sum_{n=1}^{m-1} B^{(n)\top}_{\alpha - M\beta} d_{j-1,\beta,n} \right),
\tag{7.2}
$$

see, e.g., [119] for details. For the scalar case ($r = 1$), the Equations (7.1) and (7.2) define the classical *discrete wavelet transform* (DWT). Hence, the DMWT is a natural generalization of the DWT.

For a function $f \in V_J$, the main purpose of the DMWT and its inverse is to switch between the representation

$$
f = \sum_\beta m^{\frac{J}{2}} c_{J,\beta}^\top \Phi(M^J \cdot - \beta)
\tag{7.3}
$$

and the representation

$$f = \sum_\beta m^{\frac{J'}{2}} c_{J',\beta}^\top \Phi(M^{J'} \cdot -\beta)$$

$$+ \sum_{j=J'}^{J-1} \sum_{n=1}^{m-1} \sum_\beta m^{\frac{j}{2}} d_{j,\beta,n}^\top \Psi^{(n)}(M^j \cdot -\beta) \qquad (7.4)$$

with $J' < J$. The $c_{j,\beta}$, $J' \le j \le J$, determine the approximation of f in the space V_j, therefore they are also called *approximation coefficients*. Consequently, the multiwavelet coefficients $d_{j,\beta,n}$ represent the details lost when switching from the approximation of f in V_{j+1} to its approximation in V_j. From the signal processing point of view, the DMWT acts as a tree structured multi-input, multi-output (MIMO) filter bank as is sketched in Figure 7.2. Since the detail coefficients are

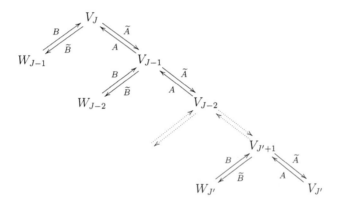

Figure 7.2: Tree structure of the DMWT and the inverse DMWT

obtained by applying the multiwavelet masks, these masks can be considered as high-pass filters whereas the low-pass filters are determined by the masks A and $\widetilde{A}$. The reader should note that the DMWT works *in place*, i.e., one step of the DMWT does not alter the amount of data since one input sequence c_j consisting of N coefficients is transformed into m sequences $c_{j-1}, (d_{j-1,\beta,1})_\beta, \ldots, (d_{j-1,\beta,m-1})_\beta$ consisting of N/m coefficients.

As we have seen in Equation (3.16), the coefficients $d_{j,\beta,n}$ can as well be determined by the inner products

$$d_{j,\beta,n} = \frac{1}{c} \langle f, m^{\frac{j}{2}} \widetilde{\Psi}^{(n)}(M^j \cdot -\beta) \rangle. \qquad (7.5)$$

Assume that f is smooth in the neighborhood of some $x_0 \in \mathbb{R}^d$, then the Taylor expansion about x_0 gives a good local approximation of f. Therefore, if the dual multiwavelet possesses a sufficiently high order of vanishing moments, most of the wavelet coefficients $d_{j,\beta,n}$ for $\beta \approx M^j x_0$ will become rather small, cf. Section 2.2. Hence, for a reasonably smooth function f the representation (7.4) will contain many more small coefficients than the representation (7.3). On the other hand, usually one does not compute the inner products (7.5) but utilizes the analysis equations (7.1) to obtain the coefficients $d_{j,\beta,n}$. Therefore, the corresponding filter banks possess an intrinsic property which resembles the vanishing moments of the wavelets, i.e., those parts of a coefficient sequence $c_{j,\beta}$ which stem from polynomial-like parts of a function are annihilated by the high-pass filters while the polynomial background or structure is preserved by the low-pass filters, see the Section 7.1.3 for a more detailed discussion of this topic. Therefore, the DMWT or DWT is extremely well suited for transform coding algorithms, since in (7.4) the information is concentrated on few large coefficients.

Another advantage of the DMWT is the low numerical complexity of $\mathcal{O}(N)$, where $N = r|\{c_{J,\beta}\}|$, see, e.g., Section 7.1 in [88]. In contrast, both the fast Fourier transform and the discrete cosine transform belong to the complexity class $\mathcal{O}(N \log N)$. A comparison between the DMWT and the DWT is somewhat more complicated. At first glance, the DMWT seems to require more arithmetic operations as the DWT. For the DMWT input data has to be vectorized, i.e., input data consisting of N elements is split into r input streams of size N/r. However, each scalar multiplication within the DWT is substituted by a matrix-vector multiplication within the DMWT. Therefore, for scalar and $r \times r$-matrix valued masks of the same support, a naive implementation of the DMWT requires r–times as many arithmetic operations as the DWT. On the other hand, if we interpret a scalar generator φ as a scaling vector Φ via (3.7), then according to Equation (6.7) the mask of Φ consist of $1/r$–times as many coefficients as the mask of φ. Taking this into account, a fair comparison reveals that the DMWT can cope with the DWT concerning computational effort. Moreover, if not only the scaling vector but also the multiwavelet is interpolating as in Chapter 4.3 and in Chapter 5.3, then by exploiting the specific structure (3.11) of the masks the DMWT can be accelerated by a factor $r/(r-1)$.

7.1.3 The Initialization Problem – Balancing

We have seen above that if we are given a proper starting sequence c_J, then the polynomial reproduction property of the generator and the vanishing moments of the wavelets are carried forward to the corresponding filter banks. However, in practice a signal is more likely to consists of the sample values of a function f than of the actual expansion coefficients. Nevertheless, a common practice is to use the sample values directly as coefficients which Strang and Nguyen [112] refer to as a *wavelet crime*. Though this is mathematically incorrect, in some applications using scalar data, it can be justified by the data acquisition process, cf. [88]. Furthermore, for the scalar case the same Taylor expansion argument as in Section 7.1.2 indicates that for large J we have $f(M^{-J}\beta) \approx c_{J,\beta}$.

For most scaling vectors Φ, however, one observes that the coefficients $c_{J,\beta}$ of a polynomial $p \in \pi_k^d$ are not necessarily the sample values of a polynomial in π_k^d and vice versa, even if Φ provides accuracy order $k+1$. Consequently, the low pass branch of the corresponding filter banks does not preserves polynomial sequences $c_{J,\beta}$ and the high pass branch does not annihilate them, cf. [83, 108]. To bypass this problem in the univariate case, Lebrun and Vetterli [83, 84] invented the notion of *balancing*. An orthonormal r-scaling vector $\Phi \in L_2(\mathbb{R})^r$ with mask A is called *balanced of order* k, if the subdivision operator $\mathcal{S}_A$ preserves the sequence $u^{(n)} := ((r\beta)^n, (r\beta+1)^n \ldots, (r\beta+r-1)^n)_{\beta\in\mathbb{Z}}$ for all $n = 0, \ldots, k-1$, i.e.,

$$\mathcal{S}_A u^{(n)} = 2^{-n} u^{(n)}. \tag{7.6}$$

Note that this is equivalent to the low-pass branch of the inverse DMWT preserving the sequence $u^{(n)}$. Let Ψ denote an orthonormal multiwavelet corresponding to Φ, and let $B \in \ell_0(\mathbb{Z}^d)^{r\times r}$ be the mask of Ψ. The unitarity of the corresponding modulation matrix $\mathcal{P}_m(z)$ implies for $\alpha \in \mathbb{Z}$

$$\sum_{\beta\in\mathbb{Z}} A_\beta A_{\beta-2\alpha}^\top = 4\,\delta_{0,\alpha}\mathbf{I}_2 \quad \text{and} \quad \sum_{\beta\in\mathbb{Z}} B_\beta A_{\beta-2\alpha}^\top = 0,$$

see Section 3.2.1 for details. Therefore, from Equation (7.6) it immediately follows that

$$\sum_{\beta\in\mathbb{Z}} A_{\beta-2\alpha} u_\beta^{(n)} = c \cdot u_\alpha^{(n)} \quad \text{and} \quad \sum_{\beta\in\mathbb{Z}} B_{\beta-2\alpha} u_\beta^{(n)} = 0,$$

where c denotes some constant. Hence, for the analysis equations (7.1) we obtain that the low-pass filter A preserves polynomial sequences whereas the high-pass filter B annihilates them. Consequently, the balancing property allows us to commit the wavelet crime. In particular, if we are given the samples $(f(\lambda_\beta))_{\beta\in\mathbb{Z}}$ of $f \in V_J$ for some equidistant nodes λ_β then, due to the definition of the sequence $u^{(n)}$, we are allowed to approximate the coefficients $c_{J,\beta}$ by the vectors $(f(\lambda_{r\beta}), \ldots, f(\lambda_{r\beta+r-1}))^\top$.

A generalization of the balancing concept to multivariate biorthogonal scaling vectors can be found in [19], see also [18]. Since only the dual masks are involved in the analysis equations (7.1) of the DMWT, it stands to reason to introduce the balancing concept for the dual scaling vector. Hence, let $(\Phi, \tilde{\Phi})$ be a pair of biorthogonal r–scaling vectors with compact support. Then $\tilde{\Phi}$ is said to be

k–*balanced* relative to $\{\xi_0, \ldots, \xi_{r-1}\} \subset \mathbb{R}^d$ if and only if

$$\int_{\mathbb{R}^d} \widetilde{\phi}_i(x)(x - \xi_i)^\mu dx = \int_{\mathbb{R}^d} \widetilde{\phi}_j(x)(x - \xi_j)^\mu dx, \quad 0 \le i,j < r, \qquad (7.7)$$

holds for all $\mu \in \mathbb{Z}_+^d$ with $|\mu| < k$. Moreover, it was shown in [18, 19] that the filter banks corresponding to balanced scaling vectors actually do possess the desired approximation properties, i.e., for all $\mu \in \mathbb{Z}_+^d$ with $|\mu| < k$ and $\alpha \in \mathbb{Z}^d$ we have

$$\sum_{\beta \in \mathbb{Z}^d} \widetilde{A}_{\beta - M\alpha} \begin{pmatrix} (\beta + \xi_0)^\mu \\ \vdots \\ (\beta + \xi_{r-1})^\mu \end{pmatrix} = \begin{pmatrix} p_\alpha(\beta + \xi_0) \\ \vdots \\ p_\alpha(\beta + \xi_{r-1}) \end{pmatrix},$$

where $p_\alpha \in \pi_{|\alpha|}^d$, and

$$\sum_{\beta \in \mathbb{Z}^d} \widetilde{B}_{\beta - M\alpha} \begin{pmatrix} (\beta + \xi_0)^\mu \\ \vdots \\ (\beta + \xi_{r-1})^\mu \end{pmatrix} = 0.$$

Hence, again, the low-pass branch of the analysis filter bank preserves polynomial sequences and the high-pass branch annihilates them.

In [18, 19], also the following handy and, moreover, implementable condition for $\widetilde{\Phi}$ being balanced is stated.

Theorem 7.1.1. *Let Φ be a compactly supported r–scaling vector with mask $A \in \ell_0(\mathbb{Z}^d)^{r \times r}$. Furthermore, let A satisfy the sum rules of order k with vectors $y_\mu = (y_\mu^1, \ldots, y_\mu^r)^\top$, $|\mu| < k$. A compactly supported dual r–scaling vector $\widetilde{\Phi}$ is K–balanced relative to $\{\xi_0, \ldots, \xi_{r-1}\} \subset \mathbb{R}^d$, $K \le k$, if and only if*

$$y_\mu^i = \sum_{\nu \le \mu} \frac{1}{\nu!} (\xi_i - \xi_0)^\nu y_{\mu - \nu}^1, \quad 1 \le i < r,$$

holds for all $\mu \in \mathbb{Z}_+^d$ with $|\mu| < K$.

In the sequel, we show that each dual $\widetilde{\Phi}$ of an interpolating scaling vector Φ is balanced up to the accuracy order of Φ. It has been shown in [10, 11] that the accuracy order provided by a compactly supported scaling vector with linearly independent integer translates is equivalent to the order of sum rules satisfied by its mask. Therefore, if a compactly supported interpolating m–scaling vector Φ provides accuracy of order k then its mask satisfies the sum rules of order k with vectors y_μ given by Lemma 5.1.5. A direct computation shows that the y_μ satisfy the assumptions of Theorem 7.1.1 with $\xi_{i+1} = M^{-1}\rho_i$. Thus, we obtain the following proposition.

Proposition 7.1.2. *Let Φ be a compactly supported interpolating m–scaling vector that provides accuracy of order k. Then each compactly supported dual m–scaling vector $\widetilde{\Phi}$ is k–balanced relative to the set $\{M^{-1}\rho_0, \ldots, M^{-1}\rho_{m-1}\}$.*

If Φ is orthonormal, we have $\Phi = \widetilde{\Phi}$, and thus Φ is balanced up to its order of accuracy. For the univariate case, this result can also be obtained by utilizing the factorization of the symbol in Theorem 4.1.4, see [79] for details.

The above results show that also for the multivariate (orthogonal or biortho-gonal) case, the balancing property allows us to commit the wavelet crime without having to bear any negative consequences, i.e., we are allowed to approximate the coefficients $c_{J,\beta}$ of a function $f \in V_J$ in the expansion (7.3) by sample values of f. Note that this is in perfect concordance with the sampling property induced by our interpolation condition, since (3.8) implies that

$$c_{J,\beta} = m^{-\frac{J}{2}} \left(f\left(M^{-J}(\beta + M^{-1}\rho_0)\right), \ldots, f\left(M^{-J}(\beta + M^{-1}\rho_{m-1})\right) \right)^{\top}. \quad (7.8)$$

On the other hand, as already mentioned at the beginning of Section 6.1, we do not need both the representation of f in V_J and a representation of f in some $\widetilde{V}_{\tilde{J}}$. Therefore, the balancing order of the primal scaling vector Φ is rather negligible.

7.1.4 Boundary Extension

One question that arises when applying the DMWT in signal processing is how to treat the boundaries of the signal. So far we have assumed that the signal X stems from the sample values of a function $f \in V_J$ for some $J \in \mathbb{Z}$ such that $X = f(\Lambda)$ for a finite index set Λ. Now, suppose we have a method to transmute the sample values into the coefficient sequence c_J of f. To apply the DMWT properly, we need the complete representation of f in V_J, i.e., we have to assume that $\mathrm{supp}(f) \subset \mathfrak{W}$, where $\mathfrak{W}$ denotes the convex hull of Λ. Furthermore, if f is at least continuous then $f(x) \to 0$ as x tends to the boundary of $\mathfrak{W}$. However, in practise signals often refuse to do us the favor of vanishing at their boundaries. Hence, a signal is more likely to stem from a truncations $f|_{\mathfrak{W}}$ than from a function supported on $\mathfrak{W}$. Thus, even if we know how to compute the coefficients c_J given the sample values of f, how can we obtain the coefficients corresponding to values outside of $\mathfrak{W}$?

A straightforward approach is to assume $c_{J,\beta} = 0$ outside of $\mathfrak{W}$. However, it is well known that such a *zero padding* leads to suboptimal compression results, cf. Chapter 10 in [112]. A more popular extension method is *periodic extension*, i.e., one assumes that the image is periodic. Unfortunately, this assumption is incompatible with $f \in V_J$ since a nontrivial periodic function can not be contained in $L_2(\mathbb{R}^2)$. This obstacle can be overcome by switching to *periodized multiwavelets*.

For a compactly supported r-scaling vector Φ and $x \in [0,1]^d$, define

$$\Phi_{j,\alpha}^{\mathrm{per}}(x) := \sum_{\beta \in \mathbb{Z}^d} \Phi(M^j(x+\beta) - \alpha), \quad \alpha \in \mathbb{Z}^d, \, j \in \mathbb{Z}, \qquad (7.9)$$

and

$$V_j^{\mathrm{per}} := \overline{\mathrm{span}\{e_i^\top \Phi_{j,\alpha}^{\mathrm{per}} \mid \alpha \in \mathbb{Z}^d, \, 0 \leq i < r\}}, \qquad (7.10)$$

where e_i denotes the ith unit vector in $\mathbb{R}^r$. It has been shown in Section 9.3 in
[37] that in the univariate scalar case the V_j^{per} define a multiresolution analysis for
$L_2([0,1]^d)$. Furthermore, if $\widetilde{\Phi}$ is dual to Φ, then its periodized version $\widetilde{\Phi}^{\mathrm{per}}$ generates
a dual MRA via (7.10) and the primal and dual complement spaces W_j^{per} and
$\widetilde{W}_j^{\mathrm{per}}$ are spanned by the periodized primal and dual multiwavelets, respectively.
In addition, all orthogonality relations are carried over to the periodic case. Since
all these properties are directly inherited from their non-periodic counterparts, we
immediately obtain the extension of these results to our general setting.

Also for the periodic case a discrete multiwavelet transform can be defined.
Let $f \in V_J^{\mathrm{per}}$ for some J. Similar to Equation (3.16), we have for $j < J$

$$
\begin{aligned}
c_{j,\alpha}^{\mathrm{per}} &:= \frac{1}{c} \big\langle f, m^{\frac{j}{2}} \widetilde{\Phi}_{j,\alpha}^{\mathrm{per}} \big\rangle_{L_2([0,1]^d)} \\
&= \frac{1}{c} \sum_{\beta \in \mathbb{Z}^d} m^{\frac{j}{2}} \big\langle f, \widetilde{\Phi}(M^j(\cdot + \beta) - \alpha) \big\rangle_{L_2([0,1]^d)}, \qquad (7.11)
\end{aligned}
$$

where the constant c is determined by Equation (3.17). With the mask $\widetilde{A} \in
\ell_0(\mathbb{Z}^d)^{r \times r}$ of $\widetilde{\Phi}$ the refinement equation yields

$$
\begin{aligned}
c_{j,\alpha}^{\mathrm{per}} &= \frac{1}{c} \sum_{\beta, \gamma \in \mathbb{Z}^d} m^{\frac{j}{2}} \widetilde{A}_\gamma \big\langle f, \widetilde{\Phi}(M^{j+1}(\cdot + \beta) - M\alpha - \gamma) \big\rangle_{L_2([0,1]^d)} \\
&= \frac{1}{c\sqrt{m}} \sum_{\gamma \in \mathbb{Z}^d} \widetilde{A}_{\gamma - M\alpha} \big\langle f, m^{\frac{j+1}{2}} \widetilde{\Phi}_{j+1,\gamma}^{\mathrm{per}} \big\rangle_{L_2([0,1]^d)} \\
&= \frac{1}{\sqrt{m}} \sum_{\gamma \in \mathbb{Z}^d} \widetilde{A}_{\gamma - M\alpha} c_{j+1,\gamma}^{\mathrm{per}}.
\end{aligned}
$$

Thus, we obtain the periodized analogon to the lowpass part of the analysis equation (7.1). Periodized versions of the highpass part of (7.1) and the synthesis
equation (7.2) can be obtained analogously. On the other hand, Equation (7.11)
implies

$$
\begin{aligned}
c_{J,\alpha}^{\mathrm{per}} &= \frac{m^{\frac{J}{2}}}{c} \sum_{\beta \in \mathbb{Z}^d} \int_{[0,1]^d} f(x) \widetilde{\Phi}(M^J(x+\beta) - \alpha) \, dx \\
&= \frac{1}{c} \big\langle f^{\mathrm{per}}, m^{\frac{J}{2}} \widetilde{\Phi}(M^J \cdot - \alpha) \big\rangle_{L_2(\mathbb{R}^d)},
\end{aligned}
$$

where f^{per} denotes the periodization of f, i.e., for $x \in [0,1]^d$ and $\beta \in \mathbb{Z}^d$ we have $f^{\mathrm{per}}(x+\beta) := f(x)$. Hence, it turns out that the periodic DMWT coincides with its non-periodic counterpart using periodized coefficient sequences. For a signal X which stems from a function $f|_{\mathfrak{W}}$ such that $\mathfrak{W}$ is an arbitrary rectangular or cubic window, the generalization of the above scheme to $\mathfrak{W}$–periodic functions is straightforward.

7.2 Explicit Image Compression

In this section, we intend to find some answers to the questions (Q1)–(Q3) posed at the beginning of this chapter by studying the performance of our (multi-)wavelets in digital image compression. First of all, we have to make some assumptions about the considered data. In general, digital images consists of real valued data $(X_\lambda)_{\lambda \in \Lambda}$ defined on a finite section Λ of some square lattice in $\mathbb{R}^2$. Each data point X_λ is called a *pixel*. Here we assume that $\Lambda \subset \mathbb{Z}^2$ and that the convex hull W of Λ has the form $W = 2^n[0,1]^2$ for some $n \in \mathbb{Z}_+$. This is due to the fact that all the examples of scaling vectors or functions constructed in this work are refinable with respect to a scaling matrix M with $m = |\det(M)| = 2$ or $M = 2$. Since in each step of the DMWT (7.1) the number of approximation coefficients is reduced by a factor $1/m$, the above choice ensures that we can compute several steps of the DMWT without having to split a pixel.

Since we focus on two-dimensional data, all the univariate (multi-)wavelets and generators have to be extended to bivariate functions. This can be performed by means of a tensor product approach as follows, cf. Chapter 7.7 in [88]. Let Φ be a univariate scaling vector, then the mapping

$$\Phi_\times : \mathbb{R}^2 \ni (x,y) \to \Phi(x) \otimes \Phi(y)^\top$$

defines a matrix valued function. By rearranging the entries of $\Phi_\times$ columnwise into a vector, we obtain a scaling vector of length r^2 which is refinable with respect to the scaling matrix $M_2 := 2\mathbf{I}_2$. The corresponding multiwavelets are given by vectorizing the matrix valued functions

$$\begin{aligned}
\Psi_\times^{(1)}(x,y) &:= \Phi(x) \otimes \Psi(y)^\top, \\
\Psi_\times^{(2)}(x,y) &:= \Psi(x) \otimes \Phi(y)^\top, \\
\Psi_\times^{(3)}(x,y) &:= \Psi(x) \otimes \Psi(y)^\top.
\end{aligned}$$

The dual functions are defined analogously. Although the vector interpretation of these functions leads to a bivariate DMWT by means of Section 7.1.2, for practical reasons it is more convenient to work with the matrix form given above. Since all these functions are separable, also the corresponding bivariate DMWT can be decomposed into products of univariate DMWTs. In particular, applying such a bivariate DMWT to an image X is equivalent to applying the corresponding univariate DMWT first to the columns and then to the rows of X.

7.2.1 Selection of Wavelets

To classify the applicability of our multiwavelets in image compression, we have to compare their performance with that of some well-established scalar wavelets. First of all, we choose the symmetric biorthogonal 9-7 wavelet pair constructed in [6]. This univariate wavelet pair can be considered as a very tough competitor since it is the basis of the lossy compression algorithm within the JPEG2000 standard, cf. [2]. The mask of the primal generator φ_7 consists of 7 nonvanishing coefficients, and we have $\mathfrak{s}(\varphi_7) = 2.12$. For the dual generator $\widetilde{\varphi}_9$, the critical Sobolev exponent is $\mathfrak{s}(\widetilde{\varphi}_9) = 1.41$, and its mask consists of 9 coefficients. Both, the primal and the dual wavelet have 4 vanishing moments. In the following, we will denote this generator/wavelet combination by $b_{9,7}$. In addition, we will denote by $b_{7,9}$ the same function set where the roles of primal and dual functions are interchanged.

Another class of reference wavelets is given by the univariate orthonormal Daubechies wavelets with vanishing moment order n, cf. [36]. The masks of the corresponding scaling functions φ_n consist of $2n$ coefficients. Here, we focus on the case $n = 3$ and $n = 4$. The regularity of the generators φ_n is $\mathfrak{s}(\varphi_3) = 1.42$ and $\mathfrak{s}(\varphi_4) = 1.78$. For conciseness, we denote these generator/wavelet combinations by d_3 and d_4, respectively.

notation	description	source
$b_{9,7}$, $b_{7,9}$	univariate symmetric biorthogonal wavelet pair	[6]
d_n	univariate orthonormal Daubechies wavelet of order n	[36]
O_n^t	orthonormal interpolating multiwavelet with index n, $t \in \{2, p, q\}$	Chap. 4/5
S_n^t	multivariate biorthogonal multiwavelet pair with primal index n, $t \in \{q, s\}$	Chap. 6
s_n^t	multivariate biorthogonal wavelet pair with primal index n, $t \in \{q, s\}$	Chap. 6

Table 7.1: Notation for the selected (multi-)wavelets

For the performance comparison, we choose those of our multiwavelets which possess properties similar to those of the reference wavelets, i.e., we focus on multiwavelets which are at least continuous and at most continuously differentiable, and which possess 3 or 4 vanishing moments. Therefore, from the univariate multiwavelets constructed in Chapter 4, we choose the interpolating multiwavelets corresponding to the scaling vectors Φ_4 and Φ_6 and denote these generator/multiwavelet pairs by O_4^2 and O_6^2, respectively. In addition, we employ the wavelets corresponding to the multivariate orthonormal scaling vectors Φ_3 and Φ_5 of Chapter 5 for quincunx and box–spline dilation. These generator/multiwavelet pairs shall be denoted by $O_3^{p/q}$ and $O_5^{p/q}$, respectively, where the index p or q indicates the scaling matrix M_p or M_q. From our symmetric (multi-)wavelets constructed in Chapter 6, we choose the scalar as well as the vector valued biorthogonal wavelet pairs with primal parameter $n = 2$ and $n = 3$ for both dilation matrices. At first glance, the choice of $n = 3$ seems to be unfair due to the higher number of vanishing moments and the smoothness of the primal functions. On the other hand, in particular the dual functions have a much larger support than the other contestants which might bear some disadvantage. For the symmetric examples, the scalar generator/multiwavelet combinations are denoted by $s_2^{q/s}$ and $s_3^{q/s}$, and for the vector valued functions we use the notation $S_2^{q/s}$ and $S_3^{q/s}$. Again, the index q or s determines the scaling matrix. For the reader's convenience, we have summarized the notation in Table 7.1.

7.2.2 Fundamental DMWT Coding Scheme

We have seen in Section 7.1 that a DMWT based transform coding scheme essentially consists of four steps. First of all, according to Section 7.1.3, image data has to be preprocessed to obtain appropriate initialization data for the discrete wavelet or multiwavelet transform. Then, taking care of a suitable boundary handling, the DMWT is applied, followed by quantization and encoding.

In our setting, we are more interested in a performance comparison than in the actual compression results. Therefore, we may omit the encoding step since it is independent of the used wavelet transform and does not alter the data. Also our quantization scheme differs from the highly advanced quantization schemes used in real world applications. As we have seen in Section 7.1.1, quantization essentially results in rounding off the values of the transformed image Ξ which results in the annihilation of small coefficients. However, in most cases also the large coefficients of Ξ are modified. Therefore, from the mathematical point of view, image compression via (multi-)wavelet transform coding can be considered as a perturbed N-term approximation, where N corresponds to the number of the remaining large coefficients, cf. [45]. Consequently, for our comparison, we may also use a true N-term approximation for quantization as follows.

We assume that $X = f(\Lambda)$ for a function $f \in V_J^{\text{per}}$ for some J. Since $\Lambda = \mathbb{Z}^2 \cap$ $\mathfrak{W}$, Equation (7.8) suggests to choose $J = -1$ for the vector case and $J = 0$ for the scalar case. Note that the choice of J is completely arbitrary, it does only describe the behaviour of f between the sample values, i.e., data between the pixels we in general do not know. Hence, the choice of J can be taken as a regularity assumption on f. As a consequence, for our interpolating generators we can obtain the starting sequences utilizing Equation (7.8) for the $\mathfrak{W}$–periodized version of f. For the non-interpolating scalar generators, we choose $c_{0,\beta} = f(\beta)$ analogously. Although this is in some cases not thoroughly correct, in the literature it has proven to be adequate. As the next step of our scheme, we apply as many steps of the discrete (multi-) wavelet transform to X as possible. This means that the number of approximation coefficients at the lowest level J' is only slightly larger than the number of filter coefficients. Afterwards, we threshold the detail coefficients, i.e., depending on the desired compression rate we keep the N' largest wavelet coefficients and set all other wavelet coefficients to zero. The approximation coefficients $c_{J'}$ remain untouched. Finally, we apply the inverse DMWT followed by a postprocessing step which is the inverse of the preprocessing step. Thus, we obtain an N-term approximation of f or X, respectively, where N equals the number of the remaining wavelet coefficients plus the number of approximation coefficients on the level J' (taking into account that these coefficients are vector valued). This compression scheme is sketched in Figure 7.3.

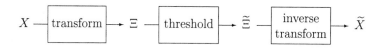

Figure 7.3: Simple DMWT compression scheme

The reader should be aware of the fact that the above compression scheme is of rudimentary nature. However, it is well suited for comparing the applicability of certain wavelets or multiwavelets in image compression, although we do not take into account the vector structure of multiwavelets. Hence, using more sophisticated compression algorithms which exploit this vector structure, the performance of the multiwavelets might be enhanced.

7.2.3   Numerical Results

In the sequel, we apply our simplified compression scheme to several 8-bit grayscale images, each consisting of 512×512 pixels. This image size allows us to compute 10 steps of the DMWT for the nonseparable (multi-)wavelets and 5 steps for the wavelets obtained by tensor products. For all images, we consider two distinct rates of compression. First of all, we produce strongly perturbed reconstructions by throwing away 99% of all wavelet coefficients which corresponds to a compression rate of $R = 1\%$. This enables us to study the typical artifacts induced by the different kinds of wavelets. In addition, we use a compression rate of $R = 10\%$, i.e., thresholding results in keeping 10% of the wavelet coefficients. This medium-sized compression rate should provide an insight into the average compression performance of our wavelets by means of the subjective distortion measures mse and mad.

Portrait Images

(a) 'barbara'   (b) 'lena'

Figure 7.4: Portrait test images

Our first class of test images are portrait images which, in general, reveal a mixture of rather long curved edges and smooth surfaces. Two typical examples of this class are the well-known test images 'barbara' and 'lena' depicted in Figure 7.4. These data sets can be obtained from various sources, e.g., from the USC-SIPI Image Database [35].

The distortion rates of the reconstructed images are shown in Table 7.2. Un-surprisingly, the best results by means of the objective measures are obtained by the JPEG2000 wavelet $b_{9,7}$. Nevertheless, also the multiwavelets O_4^2 and O_6^2 show very good results. They even outperform the Daubechies wavelets d_3 and d_4 which have a smaller support and are only slightly less regular than these univariate multiwavelets. However, for the lower compression rate $R = 10\%$, the second best

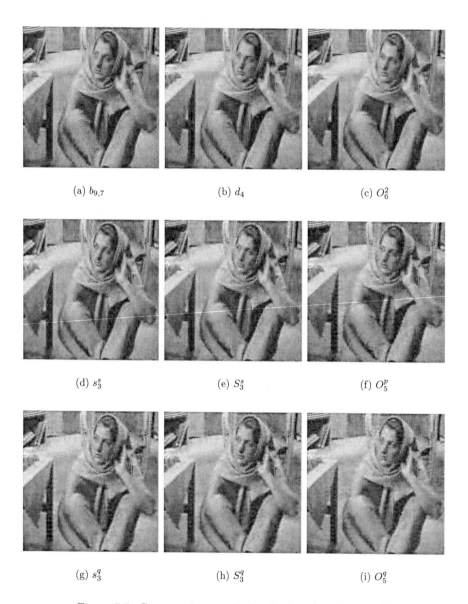

(a) $b_{9,7}$ (b) d_4 (c) O_6^2

(d) s_3^s (e) S_3^s (f) O_5^p

(g) s_3^q (h) S_3^q (i) O_5^q

Figure 7.5: Compression results for 'barbara' with $R = 1\%$

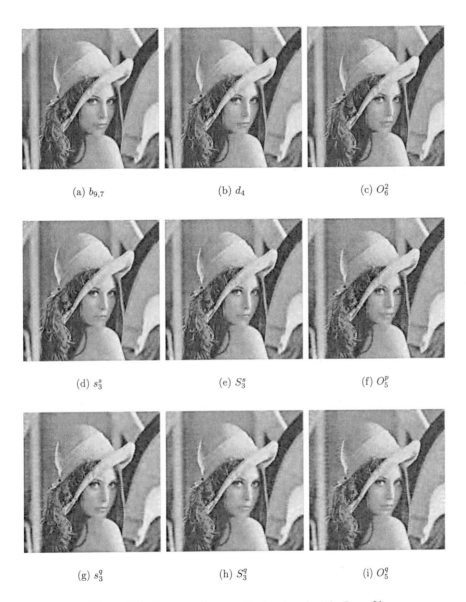

Figure 7.6: Compression results for 'lena' with $R = 1\%$

	barbara				lena			
	$R = 1\%$		$R = 10\%$		$R = 1\%$		$R = 10\%$	
	mse	mad	mse	mad	mse	mad	mse	mad
$b_{9,7}$	261.7	11.08	32.52	4.158	91.4	6.577	9.96	2.436
$b_{7,9}$	284.5	11.65	43.97	4.823	116.5	7.456	12.84	2.740
d_3	274.9	11.45	40.87	4.633	109.7	7.140	11.50	2.605
d_4	271.9	11.39	37.41	4.469	103.2	6.977	11.22	2.573
O_4^2	262.2	11.16	36.39	4.418	96.9	6.759	11.07	2.562
O_6^2	261.1	11.17	35.60	4.386	97.3	6.823	11.05	2.561
s_2^s	302.3	11.94	51.11	5.092	107.1	7.150	12.52	2.677
S_2^s	313.4	12.00	67.89	5.585	114.6	7.268	13.50	2.727
s_3^s	282.1	11.47	33.77	4.253	97.7	6.828	10.99	2.541
S_3^s	293.7	11.70	48.21	4.924	102.4	6.950	11.88	2.615
O_3^p	295.4	11.81	69.58	5.898	119.6	7.482	13.55	2.799
O_5^p	291.7	11.72	67.41	5.823	113.5	7.321	12.90	2.743
O_3^q	294.3	11.80	68.42	5.859	119.8	7.497	13.49	2.798
O_5^q	294.1	11.77	67.10	5.810	120.5	7.556	13.09	2.759
s_2^q	300.2	11.76	70.71	5.643	115.2	7.321	12.62	2.668
S_2^q	312.8	12.00	81.44	5.996	119.8	7.475	13.12	2.710
s_3^q	300.9	11.84	68.29	5.614	117.0	7.397	12.92	2.708
S_3^q	296.9	11.68	68.90	5.635	115.3	7.355	12.97	2.698

Table 7.2: Compression results for the portrait images

results are obtained by our multivariate symmetric wavelet s_3^s. Unfortunately, we observe the other nonseparable wavelets and multiwavelets can not cope with the reference wavelet $b_{9,7}$, though for the 'lena' image they perform at least similar to the twisted reference wavelet $b_{7,9}$. As we have expected, it turns out that for almost all classes of wavelets in the test those representatives with the higher order of vanishing moments and better regularity properties perform best. This should explain the gap between the results of s_2^s and s_3^s or S_2^s and S_3^s, respectively. However, we observe that in most cases the s_2^q wavelet performs better than the smoother s_3^q wavelet which may be due to the larger support of the latter.

One reason for the varying performance of the distinct wavelet classes becomes evident when looking at the reconstructed images in Figures 7.5 and 7.6. All wavelets and multiwavelets corresponding to the scaling matrices M_q and M_p induce more or less strong artifacts along high contrast edges which resemble the checkerboard structure of the quincunx grid. In contrast, the tensor product wavelets show shadow-like artifacts which run parallel to these edges and thus affect a much smaller area. The artifacts induced by the wavelets corresponding to M_s are somewhere in between, reminding of the row structure of the grid related to M_s. In the literature, the artifacts along sharp edges are called *pseudo-Gibbs phenomena*. They stem from missing wavelets bearing a significant high frequency content, cf. [27] and Chapter 2 in [112]. Now, for the 'barbara' image with its many tiny parallel stripes, the artifacts induced by the tensor product wavelets and multiwavelets do perfectly fit into these fine structures. Thus, this particular image provides an advantage for the tensor product wavelets over the M_q and M_p related wavelets. However, within the reconstructions of the 'lena' image in Figure 7.6 the artifacts induced by the quincunx wavelets are much less striking. For example, the reconstruction (h) obtained by S_3^q shows nice smooth areas but rather blurred edges. In contrast, the reconstruction (b) obtained by d_4 shows nice crisp edges but sharp artifacts which disturbingly extend into smoother areas of the image. So, choosing the more pleasing reconstruction becomes a matter of personal taste. A similar observation can be made for our objectively best performing wavelets and multiwavelets, since is hard to tell which of the reconstructions (a), (c), and (d) in Figures 7.5 and 7.6 is visually the best. Therefore, as already mentioned above, we should not rely too much on the objective measures for high compression rates.

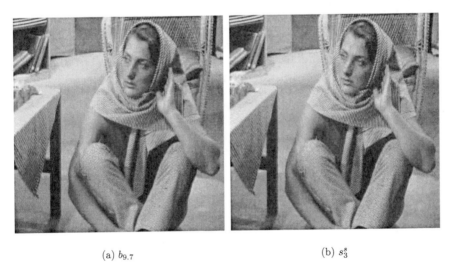

(a) $b_{9.7}$ (b) s_3^s

Figure 7.7: Compression results for 'barbara' with $R = 10\%$

In Figure 7.7 the reconstructions of the 'barbara' image obtained by the $b_{9,7}$ and the s_3^s wavelets with a compression rate $R = 10\%$ are shown. As stated above, this image contains a kind of very fine structures which gives an edge to the tensor product wavelets. Nevertheless, not only the reconstruction corresponding to the $b_{9,7}$ wavelet but also the one obtained by the s_3^s wavelet look rather close to the original picture. Even the texture of the cloth is reproduced very well. However, under close examination both reconstrcutions show some slight artifacts. Yet, to decide which artifacts are more disturbing is in the eye of the beholder.

Texture Images

The second class of images we consider consists of pictures of textures which in general thoroughly consist of fine structures without larger edges. Again, we employ two test images called 'romanesco' and 'cereals' which are shown in Figure 7.8. The 'romanesco' image indeed is a close-up photograph of a Romanesco broccoli taken from an article on fractal food by John Walker [117]. The 'cereals' image is a close-up photograph of the author's breakfast taken on one morning before writing these lines.

In Table 7.3 the distortion rates of the reconstructed images are shown. It turns out that almost all nonseparable (multi-)wavelets perform very well. In particular the M_q related symmetric wavelets and multiwavelets show pretty good results which are close to those of the $b_{9,7}$ wavelet and are superior to those obtained by the Daubechies wavelets. In addition, also the separable orthonormal multiwavelets show a very good performance. Similar to the portrait case, also for the texture images the smoother wavelets with more vanishing moments reveal better results. But again, for the symmetric quincunx wavelets s_2^q and s_3^q this is not the case. However, in spite of the fact that this time the difference is not very significant, both wavelets show a good overall performance. Quite remarkably, the worst results for the texture images are obtained by the twisted brother of the $b_{7,9}$ wavelet. Though it consists of the same functions as the $b_{9,7}$ wavelet and, moreover, the primal and the dual functions have very similar properties, the $b_{7,9}$ wavelet seems to be far less suitable for image compression purposes. This shows that in image compression DMWT or DWT based algorithms respond very sensitively to the choice of the analysis and the synthesis (multi-)wavelets.

Although the $b_{9,7}$ wavelet yields the best results in terms of the objective distortion measures, the visual performance is much less appealing. In Figures 7.9 and 7.10 some reconstructed images for $R = 1\%$ are shown. All reconstructions obtained by the separable (multi-)wavelets show similar box-like artifacts. These artifacts stem from the preferred direction phenomenon and result in somewhat grainy looking reconstructed images. This is particularly disturbing for the 'romanesco' image where the highly detailed areas, i.e., the small Romanesco florets, appear rather unnatural. In contrast, the M_q related wavelets and multiwavelets lead to much more natural looking reconstructions. Especially the symmetric

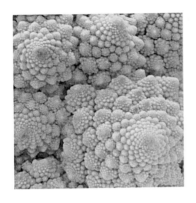

(a) 'romanesco'

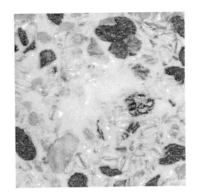

(b) 'cereals'

Figure 7.8: Texture test images

	romanesco				cereals			
	$R = 1\%$		$R = 10\%$		$R = 1\%$		$R = 10\%$	
	mse	mad	mse	mad	mse	mad	mse	mad
$b_{9,7}$	252.9	12.43	31.19	4.378	147.8	8.158	18.01	3.015
$b_{7,9}$	290.8	13.35	46.40	5.382	170.9	8.851	24.77	3.549
d_3	269.9	12.88	39.33	4.952	162.6	8.616	21.36	3.286
d_4	264.1	12.71	37.04	4.797	158.1	8.496	21.54	3.323
O_4^2	259.0	12.61	35.56	4.698	152.5	8.329	20.65	3.259
O_6^2	260.1	12.63	35.02	4.666	151.5	8.295	20.82	3.279
s_2^s	302.3	13.65	40.32	5.002	167.9	8.820	24.90	3.465
S_2^s	294.6	13.43	41.00	4.994	166.8	8.708	25.11	3.455
s_3^s	274.2	12.99	34.15	4.605	158.2	8.474	22.50	3.358
S_3^s	278.4	13.09	35.72	4.715	158.9	8.515	23.06	3.367
O_3^p	269.9	12.84	38.83	4.907	162.7	8.575	23.92	3.485
O_5^p	262.8	12.68	35.97	4.726	161.1	8.554	23.16	3.428
O_3^q	270.8	12.85	38.48	4.884	161.6	8.594	23.60	3.473
O_5^q	262.0	12.66	35.78	4.717	158.6	8.524	22.82	3.418
s_2^q	267.9	12.77	32.29	4.441	156.4	8.462	20.05	3.110
S_2^q	287.9	13.26	34.91	4.615	165.2	8.781	20.87	3.161
s_3^q	269.6	12.82	31.83	4.417	157.6	8.481	20.66	3.188
S_3^q	265.5	12.75	32.92	4.486	155.8	8.402	20.51	3.146

Table 7.3: Compression results for the texture images

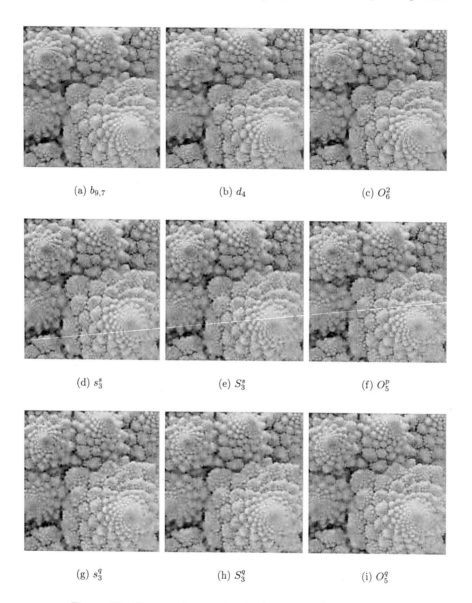

(a) $b_{9,7}$ (b) d_4 (c) O_6^2

(d) s_3^s (e) S_3^s (f) O_5^p

(g) s_3^q (h) S_3^q (i) O_5^q

Figure 7.9: Compression results for 'romanesco' with $R = 1\%$

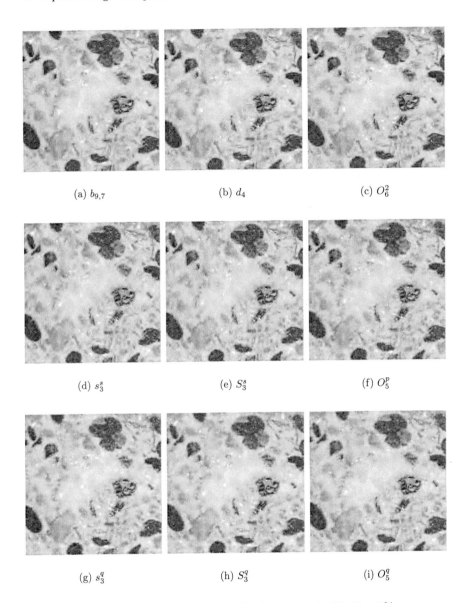

Figure 7.10: Compression results for 'cereals' with $R = 1\%$

wavelets s_3^q and S_3^q resolve the fine details very good, since the symmetry of these wavelets matches the structure of the details well. Therefore, the best visual performance is obtained by applying the symmetric quincunx (multi-)wavelets.

(a) $b_{9,7}$ (b) s_3^q

Figure 7.11: Compression results for 'romanesco' with $R = 10\%$

Yet less striking, the above observations can be confirmed for smaller compression rates as well. A close examination of Figure 7.11 reveals that also for $R = 10\%$ the reconstruction obtained by the $b_{9,7}$ wavelet suffers from the directional sensitivity. Hence, although the $b_{9,7}$ wavelet leads to better objective distortion rates, the s_3^q wavelet yields a better visual performance.

Concluding Remarks

Altogether, the numerical results obtained above allow us to give an at least partial answer to the questions (Q1)–(Q3) stated at the beginning of this chapter. First of all, we can give a positiv answer to question (Q1). In almost all cases, the univariate orthonormal multiwavelets perform very well. Though they do not show the overall best results, they can cope with the $b_{9,7}$ wavelets, and they usually outperform the Daubechies wavelets.

The answers to the other questions are far less unambiguous. We observe that the directional sensitivity of the tensor product (multi-)wavelets can have a synergetic effect, as for the strong horizontal and vertical edges in the portrait images and, moreover, for the parallel stripe textures within the 'barbara' image. There, the artifacts produced by the separable wavelets are less striking than those obtained by their nonseparable relatives. On the other hand, for images without such a strong horizontal or vertical alignment, like our texture images, the nonseparable approach can lead to far better results by means of visual quality. Thus, the choice of a proper wavelet for signal processing purposes should be made according to the type of signals considered.

The comparison between scalar and vector valued wavelets is even more difficile. Even though either one or the other of these two wavelet types may lead to better distortion rates, their visual performance is almost identical. However, we should have in mind that our simple compression scheme does not exploit the vector structure of the multiwavelets in any sense. Still, up to the author's knowledge, no sound multiwavelet-specific compression methods can be found in the known literature yet. Perhaps, developing and using a more sophisticated compression scheme could shift the results for the benefit of the multiwavelets.

Chapter 8

Summary and Perspectives

In the present work, we have studied the topic of multivariate multiwavelets associated to interpolating scaling vectors. Our main aims are formulated in the tasks (T1)–(T3) within the Introduction. In the following, we summarize our results and discuss how these tasks have been solved.

First of all, we have designed several systematic algorithms for the construction of interpolating scaling vectors. The first method given in Chapter 4 leads to univariate orthonormal interpolating scaling vectors with compact support. This approach does not only allow us to reproduce but also to improve the results obtained in [109]. The method presented in Chapter 5 is a multivariate analogon of our univariate approach which can be used to construct compactly supported orthonormal interpolating scaling vectors for scaling matrices with determinant ± 2 in arbitrary dimensions. Up to the author's knowledge, the examples provided in Chapter 5 are the first interpolating scaling vectors in $L_2(\mathbb{R}^2)$ with orthogonal integer translates. Finally, in Chapter 6, we obtain an algorithm for the construction of biorthogonal pairs of symmetric scaling vectors in $L_2(\mathbb{R}^d)$ for arbitrary scaling matrices where the primal scaling vector is interpolating. Again, the examples provided in this chapter are the first ones of their kind throughout the known literature. Moreover, these results do always possess better properties than their scalar relatives in terms of regularity per support size. Thus, all the restrictions of the scalar setting can be overcome by the vector concept. Furthermore, all three approaches have in common that they contain a postprocessing step which, by fixing all other properties, leads to a maximum Sobolev regularity. Hence, our examples are not only some isolated solutions but are optimal ones in a larger context. Hence, the task (T1) is completely and thoroughly solved.

In addition to the algorithms for the construction of scaling vectors which, by the way, may be useful for the field of vector subdivision as well, we give a solution to the problem of finding some corresponding multiwavelets. For the more general biorthogonal case, we derive a method which enables us to compute canonical multiwavelets for arbitrary dual pairs of scaling vectors whenever their masks are finitely supported and the primal scaling vector is interpolating. Hence, we obtain an at least partial solution to an in general unsolved problem. Moreover, by employing an additional interpolation property for the orthogonal case, we show that the symbols of suitable multiwavelets are obtained by a simple modification of the symbols of the corresponding scaling vector. Thus, also for the questions within (T2) we have found a complete and positive answer.

The topic (T3) leads to somewhat less unambiguous statements. First of all, as we have shown in Section 7.1.3, all interpolating scaling vectors are well-suited for application purposes since they are automatically balanced. Furthermore, the compression results obtained in Chapter 7 show that our univariate multiwavelets perform very well in image processing algorithms. However, for our multivariate examples, the compression results seem to depend on the properties of the data. Therefore, using prior knowledge about the data, our multivariate multiwavelets can be put to good use in image compression as well. Nevertheless, the reader should be aware of the fact that our construction methods are intended to fathom the potential of the vector approach in terms of regularity per support size. Hence, they do not aim at finding multiwavelets for a specific application. Especially for the biorthogonal setting, our approach may lead to suboptimal results for signal processing purposes. As we observe from the performance gap between the $b_{9,7}$ and the $b_{7,9}$ wavelets, the image compression results are vastly influenced by the properties of the primal and dual wavelets. Thus, it seems to be very important to equilibrate the properties of primal and dual functions.

This observation leads us to one possible future prospect. It seems to be worthwhile to modify our method for the construction of biorthogonal multiwavelets to better match the requirements of signal or image processing algorithms. Then some very interesting questions arise, e.g., which specific property is needed on the primal or dual side, respectively. As we have seen in the preceding chapter, both components of the $b_{9,7}$ pair possess very similar properties. The only differences are slightly distinct support sizes and a small regularity gap. Hence, what are the key properties on each side in the multivariate setting? Moreover, in our approach in Chapter 6, it is crucial to construct the primal scaling vectors first since they determine the $\widetilde{y}_\mu$ and therefore the sum rules for the dual functions. Thus, is there a canonical choice of the $\widetilde{y}_\mu$ to facilitate the equilibration of the properties of primal and dual functions?

On the other hand, due to recent developments in the fields of medical imaging and materials science, algorithms for three- or more dimensional image processing are becoming more and more important. Hence, it seems to be profitable to utilize our methods to construct scaling vectors for higher dimensions. Already in the present work, while constructing some bivariate examples, we had to face some numerical problems. One of the main difficulties is to estimate the regularity of a scaling vector which involves computing the spectrum a of large matrix. Thus, in higher dimensions, these matrices are likely to become much larger, and therefore estimating the regularity will cause even more trouble. Furthermore, the optimization methods used in our examples may need an adjustment to the more complex case. Last but not least, all these algorithms require a highly sophisticated implementation which seems to be a challenging task on its own.

Finally, it is a personal desire of the author to take up the cudgels for non-separable wavelets and multiwavelets. Up to the moment, there have been only few attempts to employ this general setting in real world applications. This may be due to the following reasons. First of all, particularly in signal processing, one is mostly concerned with two-dimensional data like images or 2×1-dimensional data like movies. For this setting, powerful algorithms based on tensor product wavelets exist. So, why should one want to use a more complex setting? More-over, from the mathematical point of view, the nonseparable setting is much less accessible than the classical approach, and therefore also the implementation of the corresponding algorithms is somewhat more involved. Nevertheless, as we have seen above, nonseparable (multi-)wavelets can at least cope with the per-formance of classical wavelets in image compression. We assume that for higher dimensional data the nonseparable approach is even more proficient, since tensor product approaches suffer from the large number of wavelets. Hence, the term mul-tiscale approximation is taken ad absurdum by the coarseness of the corresponding tensor multiresolution analysis. On the other hand, it is commonly known that scaling matrices with determinant ± 2 exist for arbitrary dimensions. Therefore, it is always possible to obtain multivariate wavelet or multiwavelet bases which stem from one single mother function or function vector, respectively.

Appendix A

A Comprehensive List of Masks

In this appendix we list all masks corresponding to the scaling vectors obtained in Chapter 4 and 6. Furthermore, we give a selection of the masks obtained by our biorthogonal approach in Chapter 6. We restrict ourselves to those examples which possess the optimal properties. To reproduce the plots of our scaling vectors and to compute their critical Sobolev exponents, we provide a small software package for MATLAB which can be downloaded from:

`www.mathematik.uni-marburg.de/~dahlke/ag-numerik/research/software/`

A.1 Masks corresponding to Chapter 4

In this section the masks obtained in Chapter 4 are given explicitly. First, we list the coefficient sequences $(a_\beta)_{\beta \in [-n, n+1]}$ of the symbol entries $a_0(z)$ corresponding to the scaling vectors Φ_n for $1 \leq n \leq 8$. Then, the masks $(B_k)_{k \in \mathbb{Z}}$ of the non-interpolating multiwavelets Ψ_n are given. We use the notation

$$B_k := \begin{pmatrix} b_k^{00} & b_k^{01} \\ b_k^{10} & b_k^{11} \end{pmatrix}.$$

Coefficient sequences $(a_\beta)_{\beta\in\mathbb{Z}}$

a_β	$n=1$	$n=2$	$n=3$	$n=4$	$n=5$	$n=6$	$n=7$	$n=8$
a_{-8}								0.00030272390021519
a_{-7}							8.46474488236856e−05	0.00095176428770024
a_{-6}						−5.9847542317558e−05	8.71705538516523e−05	0.00151857269403843
a_{-5}					−0.00085475274252818	−0.00023304881297226	−0.000323778133068634	−0.000384845790624145
a_{-4}				−0.0013020833333333	−0.0024411784910232	−0.0030220063150712	−0.00308687194730986	−0.014942995429387
a_{-3}			0.0022908090901539	−0.0053699539152725	−0.0043928385490083	−0.0088609884112656	−0.00745963593193362	−0.0212438942302372
a_{-2}		0.03125	0.03108115795423	0.037841226957636	0.041014486205116	0.044816338460119	0.0429318167079586	0.0708398092279952
a_{-1}	0.22081222792228	0.24603072956898	0.24312757272538	0.26545882007915	0.26830638162131	0.27973599714156	0.273567190946869	0.31284783122487
a_0	0.9486	0.9375	0.93800652613731	0.92553881912709	0.92285384008977	0.91465166015659	0.919901620035047	0.886903323447953
a_1	−0.22081222792228	−0.24206145913796	−0.24312757272538	−0.26415673674582	−0.27001458614461	−0.28348562819413	−0.273972002515595	−0.322106899626051
a_2	0.0514	0.03125	0.030743473862691	0.040607014206242	0.041013347410232	0.049514765055002	0.0435118223416151	0.0664820220838195
a_3		−0.0039692704310182	−0.0022908090901539	0.0034168289152725	0.0069551453339558	0.014146518040279	0.00818829675564037	0.037315954447045
a_4			0.00016884204576952	−0.0026849769576362	−0.002440267455116	−0.0067807475909773	−0.00354085673595051	−0.0142459018557781
a_5				0.0006510416666667	6.5048087899631e−07	−0.0010769051302435	−8.10334356573321e−05	−0.00756352180162508
a_6					−2.2775897679288e−07	0.00087983777665897	0.000191720573972034	0.003835339472598041
a_7						−0.00022594463322777	−3.68513507849262e−06	−2.39652249074088e−05
a_8							3.57847081591257e−06	−0.000693644681281331
a_9								0.000220624818552337

Multiwavelet masks $(B_k)_{k\in\mathbb{Z}}$

k	b_k^{00}	b_k^{01}	b_k^{10}	b_k^{11}
$n=1$, $\alpha=0.9486$				
-1	−0.222428080109859	0	0.0438833587894344	0
0	0.955541632714646	−0.999424991504081	−0.188521054922329	−0.0339070251875302
1	0.188521054922329	0.0339070251875302	0.955541632714647	−0.999424991504081
2	0.0438833587894344	0	0.222428080109859	0
$n=2$				
-2	0.0311168812088972	0	−0.0049048253369451 7	0
-1	−0.244982687543562	0	0.0386156081938253	0
0	0.929817562119305	−0.999550051522028	−0.270072770146329	0.0299949079397124
1	0.270072770146329	−0.0299949079397124	0.929817562119305	−0.999550051522028
2	0.0386156081938253	0	0.244982687543562	0
3	0.00490482533694487	0	0.0311168812088972	0
$n=3$				
-3	−0.00229081095067021	0	0.000168816819622176	0
-2	0.0310811831784831	0	−0.00229046682911585	0
-1	−0.243127911254386	0	0.0307407965668045	0
0	0.938009203374459	−0.999999999939369	−0.243117243486203	−1.10118897374539e−05
1	0.243117243486203	1.10118897373165e−05	0.93800920337446	−0.99999999993937
2	0.0307407965668045	0	0.243127911254386	0
3	0.00229046682911585	0	0.0310811831784831	0
4	0.000168816819622152	0	0.00229081095067003	0
$n=4$				
-4	−0.00268328498886854	0	0.000657980382143995	0
-3	0.00536656997773707	0	−0.00131596076428805	0
-2	0.0406157113141723	0	0.00331184495802566	0
-1	−0.265360110344017	0	0.038527333851246	0
0	0.924852859252655	−0.999996658664916	−0.266548449651388	0.00258508394493862
1	0.266548449651388	−0.00258508394493878	0.924852859252655	−0.999996658664917
2	0.038527333851246	0	0.265360110344017	0
3	−0.00331184495802567	0	0.0406157113141723	0
4	−0.00131596076428804	0	−0.00536656997773707	0
5	−0.000657980382144195	0	−0.00268328498886852	0
$n=5$				
-5	6.50480918150475e−07	0	−2.27758865018356e−07	0
-4	−0.00244117863790441	0	0.000854752323034522	0
-3	0.00695514575329282	0	−0.00244026625993947	0
-2	0.0410144854502473	0	0.00439284559696321	0
-1	−0.270014593192369	0	0.041013301010756	0
0	0.922853886195691	−0.999999999999985	−0.26830622303753	−1.7184062555909e−07
1	0.26830622303753	1.71840625647408e−07	0.922853886195691	−0.999999999999985
2	0.041013301010756	0	0.270014593192369	0
3	−0.00439284559696321	0	0.0410144854502473	0
4	−0.00244026625993949	0	−0.00695514575329281	0
5	−0.000854752323034523	0	−0.00244117863790442	0
6	−2.27758865005465e−07	0	−6.50480918022666e−07	0

A.2   Masks corresponding to Chapter 5

In this section the upper right entries $(a_\beta)_{\beta\in\Lambda}$ of the masks of Φ_n^q are given. The mask entries corresponding to the box spline case, i.e., $M = M_p$, can be derived by applying Theorem 5.2.1.

$n = 1$

Structure and coefficients of $(a_\beta)_{\beta\in\Lambda}$:

β	-1	0	1
1	$\cdot$	-0.3632965053807939	0.1564659417494347
0	0.3632965053807939	0.8435340582505653	$\cdot$

$n = 2$

Structure of $(a_\beta)_{\beta\in\Lambda}$:

β	-2	-1	0	1	2
2	$\cdot$	e	$-d_2$	b_1	$\cdot$
1	$-e$	$-d_1$	$-b_0$	a_1	$-c_1$
0	d_0	b_0	a_0	$-c_0$	$-d_2$
-1	$-b_1$	a_2	c_0	$-d_1$	$\cdot$
-2	a_3	c_1	d_0	$\cdot$	$\cdot$

Coefficients of $(a_\beta)_{\beta\in\Lambda}$:

a_0	$8.432940468331447e-1$	b_0	$3.389216872059052e-1$	d_0	$4.932029262350007e-2$
a_1	$1.358816741784712e-1$	b_1	$9.752076062522183e-4$	d_1	$4.249567159311051e-2$
a_2	$1.805671110607621e-2$	c_0	$1.232795955187407e-1$	d_2	$6.824621030389556e-3$
a_3	$2.767567882307886e-3$	c_1	$1.936777548856011e-2$	e	$1.737898637156958e-2$

$n = 3$

Structure of $(a_\beta)_{\beta\in\Lambda}$:

β	-3	-2	-1	0	1	2	3
2	$\cdot$	g	e	$-d_1$	b_1	$\cdot$	$\cdot$
1	$-g$	$-e$	$-d_2$	$-b_0$	a_1	$-c_1$	$\cdot$
0	$\cdot$	d_0	b_0	a_0	$-c_0$	$-d_1$	f_0
-1	$-d_3$	$-b_1$	a_2	c_0	$-d_2$	f_1	$\cdot$
-2	$\cdot$	a_3	c_1	d_0	$-f_1$	g	$-h$
-3	$\cdot$	$\cdot$	$-d_3$	$-f_0$	$-g$	h	$\cdot$

Coefficients of $(a_\beta)_{\beta\in\Lambda}$:

a_0	$7.901921309392772e-1$
a_1	$1.833129840241968e-1$
a_2	$2.234434072046705e-2$
a_3	$4.150544316058975e-3$
b_0	$3.779859556235052e-1$
b_1	$4.595458738534255e-3$

c_0	$1.336983912343583e-1$
c_1	$2.758348534724065e-2$
d_0	$5.687633161686027e-2$
d_1	$3.054027595325232e-2$
d_2	$2.564456744673241e-2$
d_3	$6.914882168755445e-4$

e	$6.446281631765072e-2$
f_0	$4.595458726398507e-3$
f_1	$3.858791326069464e-3$
g	$9.699853800555625e-3$
h	$1.459557138932994e-3$

$n = 4$

Structure of $(a_\beta)_{\beta\in\Lambda}$:

β	-4	-3	-2	-1	0	1	2	3	4
4	$\cdot$	$\cdot$	j_0	$-h_1$	g_3	$-e_2$	d_6	$-b_3$	$\cdot$
3	$\cdot$	$-j_1$	$-h_0$	$-g_0$	e_1	$-d_3$	b_2	a_6	$-c_3$
2	$-j_2$	h_0	g_1	e_0	d_0	$-b_1$	$-a_3$	c_2	d_6
1	h_1	g_2	$-e_0$	$-d_2$	$-b_0$	a_2	$-c_1$	$-d_3$	$-f_2$
0	g_4	$-e_1$	$-d_1$	b_0	a_0	$-c_0$	d_0	f_0	g_3
-1	e_2	$-d_4$	b_1	a_1	c_0	$-d_2$	$-f_1$	$-g_0$	$-i_1$
-2	d_5	$-b_2$	a_4	c_1	$-d_1$	f_1	g_1	$-i_0$	j_0
-3	b_3	$-a_5$	$-c_2$	$-d_4$	$-f_0$	g_2	i_0	$-j_1$	k_0
-4	a_7	c_3	d_5	f_2	g_4	i_1	$-j_2$	$-k_0$	$\cdot$

Coefficients of $(a_\beta)_{\beta\in\Lambda}$:

a_0	$7.393508266756186e-1$
a_1	$1.590366727193650e-1$
a_2	$1.040730157264092e-1$
a_3	$4.744930389692093e-3$
a_4	$2.351787322732341e-3$
a_5	$1.792405381399017e-4$
a_6	$1.065954092630927e-4$
a_7	$5.273074443778326e-6$
b_0	$4.149641472781915e-1$
b_1	$1.073735981624842e-2$
b_2	$6.876321370624657e-4$
b_3	$1.962876903607953e-5$
c_0	$1.392174328027228e-1$
c_1	$9.560418640657620e-3$

c_2	$3.206281038156767e-4$
c_3	$1.329650173101297e-5$
d_0	$3.058761467675641e-2$
d_1	$2.278312551225702e-2$
d_2	$6.470579227153199e-3$
d_3	$1.177344005935610e-3$
d_4	$2.929922981510544e-4$
d_5	$8.693077071098538e-5$
d_6	$4.949559602948002e-5$
e_0	$1.204220172363404e-1$
e_1	$3.728234527298057e-3$
e_2	$3.417226001872880e-4$
f_0	$2.924068014056131e-3$
f_1	$2.585387372456025e-3$

f_2	$2.192775353401839e-4$
g_0	$1.714248125803879e-2$
g_1	$1.112108322068441e-2$
g_2	$4.884345308568409e-3$
g_3	$8.619588028434883e-4$
g_4	$2.750939259424788e-4$
h_0	$1.801914437578222e-2$
h_1	$1.307841879260925e-3$
i_0	$1.265558617991367e-3$
i_1	$6.940258401403749e-4$
j_0	$3.299404090514096e-3$
j_1	$3.075104180937397e-3$
j_2	$2.242999095767010e-4$
k_0	$5.658604842792553e-4$

$n = 5$

Structure of $(a_\beta)_{\beta \in \Lambda}$:

β	-5	-4	-3	-2	-1	0	1	2	3	4	5
5	$\cdot$	$-n_0$	m_1	$-k_1$	j_3	$-h_2$	g_5	$-e_3$	d_7	$-b_4$	$\cdot$
4	n_0	$-m_0$	k_0	$-j_0$	h_1	$-g_3$	e_2	$-d_5$	$-b_3$	a_7	c_4
3	m_2	$-k_0$	j_1	$-h_0$	g_2	$-e_1$	d_4	b_2	$-a_5$	$-c_3$	d_7
2	k_1	j_2	h_0	g_1	e_0	$-d_3$	b_1	$-a_3$	c_2	$-d_5$	$-f_3$
1	j_4	$-h_1$	$-g_0$	$-e_0$	$-d_1$	$-b_0$	a_1	$-c_1$	d_4	$-f_2$	g_5
0	h_2	g_4	e_1	d_0	b_0	a_0	$-c_0$	$-d_3$	f_0	$-g_3$	$-i_2$
-1	g_6	$-e_2$	$-d_2$	$-b_1$	a_2	c_0	$-d_1$	$-f_1$	g_2	$-i_1$	j_3
-2	e_3	d_6	$-b_2$	$-a_4$	c_1	d_0	f_1	g_1	$-i_0$	$-j_0$	$-l_1$
-3	d_8	b_3	a_6	$-c_2$	$-d_2$	$-f_0$	$-g_0$	i_0	j_1	l_0	m_1
-4	b_4	a_8	c_3	d_6	f_2	g_4	i_1	j_2	$-l_0$	$-m_0$	$\cdot$
-5	a_9	c_4	d_8	f_3	g_6	i_2	j_4	l_1	m_2	$\cdot$	$\cdot$

Coefficients of $(a_\beta)_{\beta \in \Lambda}$:

a_0	$7.787365342903857e-1$	e_1	$8.992853890361197e-3$	g_5	$4.876740403433029e-5$
a_1	$1.563206966265842e-1$	e_2	$4.867871825262807e-4$	g_6	$1.569912663076266e-5$
a_2	$7.749999830045678e-2$	e_3	$3.477462756715321e-5$	h_0	$1.024618153667222e-2$
a_3	$8.271155512703776e-3$	f_0	$4.955451636014475e-3$	h_1	$2.407816279003956e-3$
a_4	$4.291537021864232e-3$	f_1	$4.729118599754660e-3$	h_2	$1.352189335862617e-4$
a_5	$2.915647698692102e-4$	f_2	$1.999788847486403e-4$	i_0	$2.948892987237558e-5$
a_6	$2.550502699488863e-4$	f_3	$5.643548015985818e-6$	i_1	$2.276548845913641e-5$
a_7	$3.760716383906979e-5$	d_0	$5.468640976381089e-2$	i_2	$5.191667165275111e-6$
a_8	$4.321779481775408e-6$	d_1	$5.028215879034868e-2$	j_0	$3.835472950458153e-4$
a_9	$4.887374074329685e-8$	d_2	$3.833252785442346e-3$	j_1	$2.658082630184461e-4$
b_0	$3.654554805658362e-1$	d_3	$2.232738245573430e-3$	j_2	$6.986281217643783e-5$
b_1	$2.228105086040924e-3$	d_4	$1.658784423038543e-3$	j_3	$3.357475727011834e-5$
b_2	$1.058617395881934e-3$	d_5	$1.752875013720132e-4$	j_4	$1.430146258081304e-5$
b_3	$1.056413112561339e-4$	d_6	$1.625484670874893e-4$	k_0	$1.005097425068883e-3$
b_4	$4.586135448870520e-7$	d_7	$1.197168631220558e-5$	k_1	$9.243533118832179e-5$
c_0	$1.856456585832774e-1$	d_8	$3.722982487333346e-6$	l_0	$4.082324106247016e-6$
c_1	$1.137766276335270e-2$	g_0	$1.271177367436286e-2$	l_1	$1.181464415989627e-6$
c_2	$1.547666424679079e-3$	g_1	$1.023267728924880e-2$	m_0	$9.536737680593730e-6$
c_3	$5.507259305046014e-5$	g_2	$2.754417575852340e-3$	m_1	$6.292715888320798e-6$
c_4	$1.275804213824031e-6$	g_3	$8.987066789608154e-4$	m_2	$3.244021792272933e-6$
e_0	$8.546073131670083e-2$	g_4	$5.589189575574359e-4$	n_0	$1.727830918819457e-5$

A.3 Masks corresponding to Chapter 6

In this appendix the nontrivial parts of the masks of the most relevant scaling vectors in Section 6.2.2 are listed. Again, for simplicity of notation, we consider the scalar solutions as scaling vectors. Therefore, due to Eq. (6.7), we have to give the first rows of the corresponding masks only.

Example 1: $M = M_q$

Primal Masks

$n = 1$:

$\beta \in \Lambda_1$	Φ_1^{sc} $a_\beta^{0,1}$	Φ_1^∞ $a_\beta^{0,1}$	$a_{\beta+\rho_1}^{1,1}$
0	1/4	369/1024	143/1024

$n = 2$:

$\beta \in \Lambda_1$	Φ_2^{sc} $a_\beta^{0,1}$	Φ_2^v $a_\beta^{0,1}$	$a_{\beta+\rho_1}^{1,1}$
0	5475/16384	177/512	159/512
$(1,0)^\top$	$-867/16384$	$-27/512$	$-21/512$
$(1,1)^\top$	355/16384	5/512	11/512

$n = 3$:

$\beta \in \Lambda_1$	Φ_3^{sc} $a_\beta^{0,1}$	Φ_3^v $a_\beta^{0,1}$	$a_{\beta+\rho_1}^{1,1}$
0	89/256	7043/16384	1835/8192
$(1,0)^\top$	$-231/4096$	$-2411/65536$	$-115/4096$
$(1,1)^\top$	$-11/4096$	$-571/8192$	113/8192
$(2,0)^\top$	7/4096	$-553/65536$	11/2048
$(2,1)^\top$	21/2048	$-293/65536$	113/4096
$(2,2)^\top$	$-25/4096$	$-353/32768$	5/2048

$n = 4$:

$\beta \in \Lambda_1$	Φ_4^{sc} $a_\beta^{0,1}$	Φ_4^v $a_\beta^{0,1}$	$a_{\beta+\rho_1}^{1,1}$
0	$5819/16384$	$25671/65536$	$20201/65536$
$(1,0)^\top$	$-459/8192$	$-903/16384$	$-685/16384$
$(1,1)^\top$	$-629/16384$	$-4759/65536$	$-1893/65536$
$(2,0)^\top$	$21/8192$	$-5/16384$	$1/256$
$(2,1)^\top$	$621/16384$	$305/8192$	$181/4096$
$(2,2)^\top$	$-573/16384$	$-2101/65536$	$-2703/65536$
$(3,0)^\top$	$-43/16384$	$-29/4096$	$-23/16384$
$(3,1)^\top$	$-27/8192$	$-3/2048$	$-19/16384$
$(3,2)^\top$	$53/8192$	$35/4096$	$15/4096$
$(3,3)^\top$	$-29/16384$	$-43/65536$	$-189/65536$

Dual Masks

primal: Φ_1^{sc}

$\beta \in \Lambda_0$	$\widetilde{\Phi}_{2,2}^{sc}$ $\widetilde{a}_\beta^{0,0}$	$\widetilde{\Phi}_{2,2}^v$ $\widetilde{a}_\beta^{0,0}$	$\widetilde{a}_{\beta+\rho_1}^{1,0}$
0	$1.600107924056818e + 0$	$1.463447048560035e + 0$	$1.852368110309690e + 0$
$(1,0)^\top$	$-1.530785859804724e - 1$	$-2.458651577561153e - 1$	$-6.186321224181324e - 2$
$(1,1)^\top$	$-6.250000000000000e - 2$	$-9.684522260113693e - 2$	$-2.815477739886307e - 2$
$(2,0)^\top$	$4.686745199111852e - 2$	$2.241131796386697e - 2$	$1.195273260334195e - 2$
$(2,1)^\top$	$1.403929299023620e - 2$	$2.608735627692071e - 2$	$2.776828722043545e - 3$
$(2,2)^\top$	$-9.394433005323059e - 3$	$1.488169729498723e - 2$	$-3.199537579627432e - 3$
$\beta \in \Lambda_1$	$\widetilde{a}_\beta^{0,1}$	$\widetilde{a}_\beta^{0,1}$	$\widetilde{a}_{\beta+\rho_1}^{1,1}$
0	$3.998920759431819e - 1$	$5.365529514399645e - 1$	$1.476318896903104e - 1$
$(1,0)^\top$	$-9.373490398223705e - 2$	$-4.482263592773394e - 2$	$-2.390546520668391e - 2$
$(1,1)^\top$	$3.757773202129223e - 2$	$-5.952678917994891e - 2$	$1.279815031850973e - 2$

primal: Φ_1^{∞}

$\beta \in \Lambda_0$	$\widetilde{\Phi}_{2,2}^v$ $\widetilde{a}_\beta^{0,0}$	$\widetilde{a}_{\beta+\rho_1}^{1,0}$
0	$1.399799757734565e + 0$	$1.956178711753553e + 0$
$(1,0)^\top$	$-1.093577505501146e - 1$	$-3.690526208030116e - 2$
$(1,1)^\top$	$-1.156879524547224e - 1$	$-5.201651558494899e - 3$
$(2,0)^\top$	$1.791124034256157e - 2$	$7.608604888852144e - 3$
$(2,1)^\top$	$9.845874730273253e - 3$	$5.030187508090313e - 3$
$(2,2)^\top$	$1.645086776907467e - 2$	$-1.854934385735247e - 3$
$\beta \in \Lambda_1$	$\widetilde{a}_\beta^{0,1}$	$\widetilde{a}_{\beta+\rho_1}^{1,1}$
0	$4.163990840649086e - 1$	$7.844929923839497e - 2$
$(1,0)^\top$	$-2.485245272463827e - 2$	$-2.724199792372236e - 2$
$(1,1)^\top$	$-4.565227261661915e - 2$	$1.328288679016009e - 2$

primal: Φ_2^{sc}

	$\widetilde{\Phi}_{3,2}^{sc}$	$\widetilde{\Phi}_{3,2}^{v}$	
$\beta \in \Lambda_0$	$\widetilde{a}_\beta^{0,0}$	$\widetilde{a}_\beta^{0,0}$	$\widetilde{a}_{\beta+\rho_1}^{1,0}$
0	$1.489959004741453e+0$	$1.377640177610037e+0$	$1.868570093649113e+0$
$(1,0)^\top$	$-1.504592340501103e-1$	$-2.535438871169237e-1$	$-6.888496860165862e-2$
$(1,1)^\top$	$-6.147499859423500e-2$	$-8.238148489787550e-2$	$-3.744404927728919e-2$
$(2,0)^\top$	$9.486424762927366e-2$	$5.801682282197251e-2$	$-1.434458451547471e-3$
$(2,1)^\top$	$1.370617952505521e-2$	$4.039188537918357e-2$	$-2.224332519892439e-3$
$(2,2)^\top$	$-2.985399881463683e-2$	$8.550020327941846e-3$	$-1.684952513154365e-3$
$(3,0)^\top$	$-9.697015949889556e-3$	$3.807987439147264e-4$	$1.735556974667535e-3$
$(3,1)^\top$	$-4.843890576497218e-5$	$-3.998573281402748e-3$	$7.772324565674431e-4$
$(3,2)^\top$	$3.871945474944779e-3$	$-2.833453640255469e-3$	$-1.778492190356619e-4$
$(3,3)^\top$	$-9.281235942350252e-4$	$1.355519013104642e-3$	$-8.730318826933762e-5$
$\beta \in \Lambda_1$	$\widetilde{a}_\beta^{0,1}$	$\widetilde{a}_\beta^{0,1}$	$\widetilde{a}_{\beta+\rho_1}^{1,1}$
0	$3.553348646984411e-1$	$4.626878567340421e-1$	$1.037813036206012e-1$
$(1,0)^\top$	$-7.408486469844125e-2$	$-2.201767035819725e-2$	$1.804851000355395e-2$
$(1,1)^\top$	$4.283486469844128e-2$	$-6.256006622734213e-2$	$4.029226581985430e-3$

primal: Φ_2^{v}

	$\widetilde{\Phi}_{2,2}^{v}$	
$\beta \in \Lambda_0$	$\widetilde{a}_\beta^{0,0}$	$\widetilde{a}_{\beta+\rho_1}^{1,0}$
0	$1.399144982178193e+00$	$1.818511794595736e+00$
$(1,0)^\top$	$-2.047539926076301e-01$	$-1.044501508274394e-01$
$(1,1)^\top$	$-5.736299585011739e-02$	$-6.036174690265650e-02$
$(2,0)^\top$	$7.180021931877056e-02$	$-4.951402652602507e-03$
$(2,1)^\top$	$3.737785007318260e-02$	$-7.870243150870260e-03$
$(2,2)^\top$	$1.133823879578236e-02$	$-7.601249655432685e-03$
$(3,0)^\top$	$-1.331158391306819e-03$	$2.629650603141076e-03$
$(3,1)^\top$	$-5.144770248195471e-03$	$1.450400104696072e-03$
$(3,2)^\top$	$-1.410570914612622e-03$	$8.557896039174812e-04$
$(3,3)^\top$	$5.772400056101973e-04$	$-4.637569658378880e-04$
$\beta \in \Lambda_1$	$\widetilde{a}_\beta^{0,1}$	$\widetilde{a}_{\beta+\rho_1}^{1,1}$
0	$4.217895800803504e-01$	$1.522523359908732e-01$
$(1,0)^\top$	$-4.718950907113650e-02$	$2.893034993599408e-02$
$(1,1)^\top$	$-5.910937657448421e-02$	$2.158577877354517e-02$

primal: Φ_3^{sc}

	$\widetilde{\Phi}_{5,3}^{sc}$	$\widetilde{\Phi}_{5,3}^{v}$	
$\beta \in \Lambda_0$	$\widetilde{a}_\beta^{0,0}$	$\widetilde{a}_\beta^{0,0}$	$\widetilde{a}_{\beta+\rho_1}^{1,0}$
0	$1.447315427337890e+0$	$1.434917708020856e+0$	$1.457318747769052e+0$
$(1,0)^\top$	$-1.494398043002927e-1$	$-1.428139986885572e-1$	$-1.757594606675662e-1$
$(1,1)^\top$	$-5.187832844550799e-2$	$3.018988426543162e-4$	$-9.250132641721648e-2$
$(2,0)^\top$	$9.916756434062496e-2$	$9.077333268652309e-2$	$1.087475048523288e-1$
$(2,1)^\top$	$1.616502788076173e-2$	$4.557508757954488e-2$	$-3.159822261051633e-3$
$(2,2)^\top$	$-2.362183329687495e-2$	$8.467238139995101e-3$	$-5.768604669108011e-2$
$(3,0)^\top$	$-2.763583762499999e-2$	$-4.318260784200491e-2$	$8.749912394307897e-3$
$(3,1)^\top$	$-4.287893286767562e-3$	$-2.226586633482490e-2$	$1.425975037117529e-3$
$(3,2)^\top$	$9.076820767431633e-3$	$7.406838087778502e-3$	$6.071494325419688e-4$
$(3,3)^\top$	$-1.918720430175784e-3$	$6.224648083032844e-3$	$2.766350546015059e-3$
$(4,0)^\top$	$6.763089044921895e-4$	$5.777110821272881e-3$	$-4.413002426953758e-3$
$(4,1)^\top$	$1.906619874853513e-3$	$3.387980277251427e-3$	$1.869752932544759e-4$
$(4,2)^\top$	$-7.849469203124989e-4$	$-1.013557599140401e-3$	$8.259512870494577e-4$
$(4,3)^\top$	$-1.076315285009764e-3$	$-1.444035411099614e-3$	$-9.884310455286005e-4$
$(4,4)^\top$	$1.018997057910156e-3$	$8.414394447404391e-5$	$5.658173501448381e-4$
$(5,0)^\top$	$5.772288393554686e-4$	$5.904048785991063e-6$	$3.424583159325150e-6$
$(5,1)^\top$	$-4.820326622558596e-4$	$-4.799639953116584e-4$	$2.227740785187213e-4$
$(5,2)^\top$	$-6.484669506835955e-5$	$-1.415787832071395e-4$	$6.632985064342088e-5$
$(5,3)^\top$	$3.892003868652338e-4$	$4.127170074684915e-4$	$-1.986509989652709e-4$
$(5,4)^\top$	$-2.581000000000000e-4$	$1.519688981826937e-5$	$-1.327682400845889e-5$
$(5,5)^\top$	$5.850000000000001e-5$	$-5.618288115270778e-5$	$3.064223866117416e-5$
$\beta \in \Lambda_1$	$\widetilde{a}_\beta^{0,1}$	$\widetilde{a}_\beta^{0,1}$	$\widetilde{a}_{\beta+\rho_1}^{1,1}$
0	$3.679062652000011e-1$	$3.673714726486456e-1$	$3.726546939270025e-1$
$(1,0)^\top$	$-9.739715899999979e-2$	$-1.200828106926882e-1$	$-5.662469366105900e-2$
$(1,1)^\top$	$7.475394279999997e-2$	$-2.575813272538062e-2$	$1.115248959914140e-1$
$(2,0)^\top$	$2.245964379999995e-2$	$4.324491458560385e-2$	$-2.062607680750471e-2$
$(2,1)^\top$	$-2.618490880000000e-2$	$-1.297454702891494e-2$	$6.259038116628677e-3$
$(2,2)^\top$	$9.584640000000002e-3$	$-9.205003248059642e-3$	$5.020424382246774e-3$

primal: Φ_3^v

$\beta \in \Lambda_0$	$\widetilde{\Phi}_{5,3}^v$	
	$\widetilde{a}_\beta^{0,0}$	$\widetilde{a}_{\beta+\rho_1}^{1,0}$
0	$1.435905158157269e+0$	$1.727885226275696e+0$
$(1,0)^\top$	$-1.645496711823307e-1$	$-1.953194784762911e-1$
$(1,1)^\top$	$-3.041564819165798e-2$	$-8.560999058316465e-2$
$(2,0)^\top$	$4.903060934772241e-2$	$-2.560166185479251e-2$
$(2,1)^\top$	$-9.392102227921195e-3$	$3.890120072223180e-2$
$(2,2)^\top$	$4.411151746033415e-2$	$-2.445842541311460e-2$
$(3,0)^\top$	$-9.782508976429738e-3$	$1.132971422347406e-2$
$(3,1)^\top$	$-8.435585869282280e-3$	$-1.732231628961434e-2$
$(3,2)^\top$	$-1.756784014003271e-2$	$6.439579720562916e-3$
$(3,3)^\top$	$2.921857869606054e-3$	$-5.834494747569221e-5$
$(4,0)^\top$	$3.594250567488050e-3$	$-2.492899728881769e-5$
$(4,1)^\top$	$1.100033847582296e-4$	$1.165697043471322e-3$
$(4,2)^\top$	$1.288046291019834e-3$	$-1.238009280969662e-3$
$(4,3)^\top$	$1.001897214378523e-3$	$6.394968777380557e-4$
$(4,4)^\top$	$-5.245893293792780e-4$	$-3.235515958444042e-4$
$(5,0)^\top$	$-1.164319479827057e-3$	$1.763497017252765e-4$
$(5,1)^\top$	$7.387715108399147e-4$	$-1.921613809934466e-4$
$(5,2)^\top$	$-7.461536801662422e-4$	$5.432997984785985e-5$
$(5,3)^\top$	$2.910515258360107e-4$	$-9.947195162347106e-5$
$(5,4)^\top$	$-1.227747126128741e-4$	$9.134245078439554e-5$
$(5,4)^\top$	$6.948615475677912e-5$	$-1.704507895634243e-5$
$\beta \in \Lambda_1$	$\widetilde{a}_\beta^{0,1}$	$\widetilde{a}_{\beta+\rho_1}^{1,1}$
0	$3.455025863243808e-1$	$2.703911707545501e-1$
$(1,0)^\top$	$-2.490210949480987e-2$	$2.185848958002024e-2$
$(1,1)^\top$	$-3.869595091541100e-2$	$7.716612997634015e-4$
$(2,0)^\top$	$1.553766661550493e-2$	$6.729120883960320e-3$
$(2,1)^\top$	$1.084354693782524e-2$	$3.062158804930768e-3$
$(2,2)^\top$	$3.637256100741115e-3$	$-3.405689922203835e-3$

primal: Φ_4^{sc}

	$\widetilde{\Phi}_{7,4}^{sc}$	$\widetilde{\Phi}_{7,4}^{v}$	
$\beta \in \Lambda_0$	$\widetilde{a}_\beta^{0,0}$	$\widetilde{a}_\beta^{0,0}$	$\widetilde{a}_{\beta+\rho_1}^{1,0}$
0	$1.458911886757529e+0$	$1.396326786315009e+0$	$1.521346127569127e+0$
$(1,0)^\top$	$-1.778461112855403e-1$	$-1.986667789131478e-1$	$-1.547563314813057e-1$
$(1,1)^\top$	$-2.093876624779061e-2$	$-1.544903102142589e-2$	$-3.135088448472284e-2$
$(2,0)^\top$	$6.857354861895532e-2$	$6.255291191188503e-2$	$8.751851770230691e-2$
$(2,1)^\top$	$3.236481070927068e-2$	$3.676452480358103e-2$	$2.323270716823020e-2$
$(2,2)^\top$	$-1.717426290204790e-3$	$5.200412965743353e-3$	$-1.749197937158220e-2$
$(3,0)^\top$	$-4.596678882003101e-3$	$-3.124540661208656e-2$	$1.624706703902110e-2$
$(3,1)^\top$	$-3.133410975622854e-2$	$-4.210422075601192e-2$	$-1.429238886134079e-2$
$(3,2)^\top$	$-4.329058820641301e-3$	$-4.206266834749632e-3$	$4.320806368679196e-4$
$(3,3)^\top$	$2.081390698034639e-2$	$1.666413402923058e-2$	$1.825191628947080e-2$
$(4,0)^\top$	$1.569286498286392e-2$	$8.154998095399735e-3$	$1.025119877205654e-2$
$(4,1)^\top$	$3.895349498565826e-3$	$4.285140465989735e-3$	$7.413107823787608e-3$
$(4,2)^\top$	$-4.051827858865090e-3$	$-3.379083689063001e-3$	$-2.317967585919351e-3$
$(4,3)^\top$	$-2.839851670560841e-3$	$-4.151383008697436e-3$	$-3.925628917282185e-3$
$(4,4)^\top$	$-1.763789633477622e-3$	$-3.664961964335934e-4$	$1.005378056087144e-3$
$(5,0)^\top$	$-3.892822704883545e-3$	$-1.324791758477191e-3$	$-2.914321975227278e-3$
$(5,1)^\top$	$1.418253898066458e-4$	$-5.810155787312109e-4$	$-5.515600313668114e-4$
$(5,2)^\top$	$3.838208750554210e-4$	$3.442958521348178e-4$	$3.232051831561194e-4$
$(5,3)^\top$	$4.556301657928507e-4$	$1.160149767548107e-3$	$2.454799464058327e-4$
$(5,4)^\top$	$1.164864453559067e-3$	$3.626887278286816e-4$	$4.004012043884906e-4$
$(5,5)^\top$	$-8.987990829219059e-4$	$-4.846667609708030e-4$	$-3.783409944884670e-4$
$(6,0)^\top$	$-5.553645445971952e-5$	$-5.990299170780752e-5$	$7.946062023815245e-5$
$(6,1)^\top$	$4.566771922883294e-4$	$1.509295306249043e-4$	$4.527089197231077e-4$
$(6,2)^\top$	$-4.140111896491555e-5$	$-1.761617748317946e-4$	$8.130273144277811e-5$
$(6,3)^\top$	$-5.358995772262187e-4$	$-2.932436779286007e-4$	$-3.626874305372066e-4$
$(6,4)^\top$	$1.304494204664221e-4$	$1.352649303700866e-4$	$4.233489034190845e-5$
$(6,5)^\top$	$1.085838920218643e-4$	$5.527206749547874e-5$	$5.540956823780667e-5$
$(6,6)^\top$	$-3.207379833170564e-5$	$-1.642549248204695e-5$	$-1.768154722622150e-5$
$(7,0)^\top$	$2.433398549930865e-6$	$4.837745679320686e-5$	$-5.417270332292274e-5$
$(7,1)^\top$	$-7.516762843217175e-5$	$1.472699790675799e-6$	$-8.480845148160575e-5$
$(7,2)^\top$	$5.654627775807603e-5$	$1.906963682479146e-5$	$4.262163684047053e-5$
$(7,3)^\top$	$6.142310065938417e-5$	$6.34013181428 2471e-6$	$6.229043402323261e-5$
$(7,4)^\top$	$-5.909786200839252e-5$	$-2.926209254763072e-5$	$-3.253536435593549e-5$
$(7,5)^\top$	$1.220139551698823e-5$	$5.393340374289127e-6$	$6.276047618224489e-6$
$(7,6)^\top$	$-1.552311440337069e-7$	$-2.317529894409219e-7$	$2.553272538335563e-7$
$(7,7)^\top$	$5.301613641468597e-8$	$5.583477050759378e-7$	$-5.027820829441454e-7$
$\beta \in \Lambda_1$	$\widetilde{a}_\beta^{0,1}$	$\widetilde{a}_\beta^{0,1}$	$\widetilde{a}_{\beta+\rho_1}^{1,1}$
0	$3.535774703388350e-1$	$3.961584661555971e-1$	$3.170849596965070e-1$
$(1,0)^\top$	$-5.782201964437747e-2$	$-6.948004572752114e-2$	$-6.201468211115798e-2$
$(1,1)^\top$	$-2.789477543454291e-2$	$-3.427844444210647e-2$	$2.238986209599518e-2$
$(2,0)^\top$	$1.150915699951111e-2$	$3.688916598661446e-2$	$-1.089002623088525e-2$
$(2,1)^\top$	$1.898513583553627e-2$	$9.138665622549374e-3$	$7.613294936478802e-3$
$(2,2)^\top$	$-1.855318274669944e-2$	$-8.040870544001317e-3$	$-1.091251275369930e-2$
$(3,0)^\top$	$-8.189642068967714e-3$	$3.742815034438686e-3$	$-1.334612155359254e-2$
$(3,1)^\top$	$6.917206118384043e-3$	$6.195550835068113e-3$	$8.068925406940195e-4$
$(3,2)^\top$	$2.178053665199900e-5$	$1.022081455892033e-3$	$-8.940172229488332e-4$
$(3,3)^\top$	$2.995228893166258e-5$	$3.154471999987643e-4$	$-2.840545395502372e-4$

Example 2: $M = M_s$

Primal Masks

$n = 1:$

$\beta \in \Lambda_1$	Φ_1^{sc} $a_\beta^{0,1}$
0	$103/256$
$(0,1)^\top$	$25/512$

$n = 2:$

$\beta \in \Lambda_1$	Φ_2^{sc} $a_\beta^{0,1}$	Φ_2^v $a_\beta^{0,1}$	$a_{\beta+\rho_1}^{1,1}$
0	$111/256$	$63/128$	$17/64$
$(1,0)^\top$	$-25/256$	$-17/128$	$3/32$
$(0,1)^\top$	$7/64$	$7/32$	$31/512$
$(1,1)^\top$	0	$-1/8$	$17/512$
$(0,2)^\top$	$-23/512$	$-13/512$	$-9/128$
$(1,2)^\top$	$9/512$	$1/512$	$3/64$

$n = 3:$

$\beta \in \Lambda_1$	Φ_3^{sc} $a_\beta^{0,1}$	Φ_3^v $a_\beta^{0,1}$	$a_{\beta+\rho_1}^{1,1}$
0	$4543/8192$	$2359/4096$	$3263/8192$
$(1,0)^\top$	$-745/4096$	$-1529/8192$	$43/8192$
$(2,0)^\top$	$203/8192$	$177/8192$	$15/2048$
$(0,1)^\top$	$327/8192$	$2301/16384$	$31/2048$
$(1,1)^\top$	$157/4096$	$-1299/16384$	$211/4096$
$(2,1)^\top$	$-11/8192$	$93/16384$	$3/16384$
$(0,2)^\top$	$-375/16384$	$-15/2048$	$-725/16384$
$(1,2)^\top$	$-19/4096$	$-29/2048$	$69/4096$
$(2,2)^\top$	$-53/16384$	$-43/8192$	$11/16384$
$(0,3)^\top$	$-11/8192$	$-39/16384$	$-23/8192$
$(1,3)^\top$	$69/8192$	$45/4096$	$117/16384$
$(2,3)^\top$	$-1/512$	$-17/4096$	$1/8192$

Dual Masks

primal: Φ_1^{sc}

	$\widetilde{\Phi}_{2,2}^{sc}$	$\widetilde{\Phi}_{2,2}^{v}$	
$\beta \in \Lambda_0$	$\widetilde{a}_\beta^{0,0}$	$\widetilde{a}_\beta^{0,0}$	$\widetilde{a}_{\beta+\rho_1}^{1,0}$
0	$1.440549531452259e+0$	$1.444779063448928e+0$	$1.447189517747492e+0$
$(1,0)^\top$	$-2.190764089509072e-1$	$-2.193198643337776e-1$	$-2.183223940667431e-1$
$(2,0)^\top$	$6.064882532296319e-2$	$5.829060394175832e-2$	$5.808284705951094e-2$
$(0,1)^\top$	$-3.307652529873131e-2$	$-3.610604387268384e-2$	$-3.477143757410776e-2$
$(1,1)^\top$	$-1.669829648800684e-2$	$-1.708466194495554e-2$	$-1.653195114763771e-2$
$(2,1)^\top$	$-1.600338386411846e-4$	$9.683599913863785e-4$	$8.537676394161649e-4$
$(0,2)^\top$	$4.353723050402365e-3$	$3.910721104261113e-3$	$4.041024811857492e-3$
$(1,2)^\top$	$1.236500963460443e-3$	$1.191883314526538e-3$	$1.245858978327107e-3$
$(2,2)^\top$	$-9.403605617407395e-4$	$-7.634772376040183e-4$	$-7.746534276016387e-4$
$\beta \in \Lambda_1$	$\widetilde{a}_\beta^{0,1}$	$\widetilde{a}_\beta^{0,1}$	$\widetilde{a}_{\beta+\rho_1}^{1,1}$
0	$7.060603211166391e-1$	$6.997031503031850e-1$	$6.970314959238501e-1$
$(1,0)^\top$	$-1.554132416532994e-1$	$-1.486727665509067e-1$	$-1.482119553608431e-1$
$(0,1)^\top$	$-4.458212403612022e-2$	$-4.004578410763379e-2$	$-4.138009407342071e-2$
$(1,1)^\top$	$1.925858430445035e-2$	$1.563601382613029e-2$	$1.586490219728156e-2$

primal: Φ_2^{sc}

	$\widetilde{\Phi}_{3,2}^{sc}$	$\widetilde{\Phi}_{3,2}^{v}$	
$\beta \in \Lambda_0$	$\widetilde{a}_\beta^{0,0}$	$\widetilde{a}_\beta^{0,0}$	$\widetilde{a}_{\beta+\rho_1}^{1,0}$
0	$1.441170242241141e+0$	$1.421859276886160e+0$	$1.764362012832098e+0$
$(1,0)^\top$	$-1.550562346343914e-1$	$-1.595892336521966e-1$	$-1.856946251114907e-1$
$(2,0)^\top$	$1.114461288794294e-1$	$1.211016115569202e-1$	$-5.014975641604906e-2$
$(3,0)^\top$	$-1.291251536560865e-2$	$-8.379516347803423e-3$	$1.772587511149067e-2$
$(0,1)^\top$	$-1.056158356640418e-1$	$-2.425261921542349e-1$	$-5.874423628080448e-2$
$(1,1)^\top$	$-5.747918989364955e-2$	$-4.491744979803883e-2$	$-5.562066029634941e-2$
$(2,1)^\top$	$-1.879582167979076e-3$	$6.657559607711744e-2$	$-2.531538185959776e-2$
$(3,1)^\top$	$2.791689893649548e-3$	$-9.770050201961174e-3$	$9.331602963494031e-4$
$(0,2)^\top$	$6.907597437746077e-2$	$2.895873830031883e-2$	$2.478187566683178e-2$
$(1,2)^\top$	$1.134762223419044e-2$	$1.216356205739538e-2$	$1.686253252006832e-2$
$(2,2)^\top$	$-2.086611218873038e-2$	$-8.074941501594150e-4$	$1.280937166584112e-3$
$(3,2)^\top$	$2.324252765809557e-3$	$1.508312942604616e-3$	$-3.190657520068320e-3$
$(0,3)^\top$	$-4.357759833989537e-3$	$1.525081007135403e-2$	$-1.456640462594190e-3$
$(1,3)^\top$	$6.128099766547787e-4$	$-2.144645166284160e-3$	$2.048400650523080e-4$
$(2,3)^\top$	$2.178879916994769e-3$	$-7.625405035677013e-3$	$7.283202312970950e-4$
$(3,3)^\top$	$-6.128099766547787e-4$	$2.144645166284160e-3$	$-2.048400650523080e-4$
$\beta \in \Lambda_1$	$\widetilde{a}_\beta^{0,1}$	$\widetilde{a}_\beta^{0,1}$	$\widetilde{a}_{\beta+\rho_1}^{1,1}$
0	$6.322241573438326e-1$	$5.858062474015070e-1$	$3.184870388583356e-1$
$(1,0)^\top$	$-1.322241573438326e-1$	$-8.580624740150704e-2$	$1.815129611416644e-1$
$(0,1)^\top$	$-3.486207867191630e-2$	$1.220064805708322e-1$	$-1.165312370075352e-2$
$(1,1)^\top$	$3.486207867191630e-2$	$-1.220064805708322e-1$	$1.165312370075352e-2$

primal: Φ_2^v

$\beta \in \Lambda_0$	$\widetilde{\Phi}_{2,2}^v$	
	$\widetilde{a}_\beta^{0,0}$	$\widetilde{a}_{\beta+\rho_1}^{1,0}$
0	$1.476846649101118e+0$	$1.875366517566218e+0$
$(1,0)^\top$	$-7.147504047098890e-2$	$-9.199266653031440e-2$
$(2,0)^\top$	$1.659345856056912e-1$	$-3.332534862685921e-2$
$(3,0)^\top$	$-2.416704937276110e-2$	$-3.649423313435604e-3$
$(0,1)^\top$	$-9.413559239331361e-2$	$-1.898190047712446e-1$
$(1,1)^\top$	$-7.827371859238917e-2$	$-7.590752676264784e-2$
$(2,1)^\top$	$-2.329048017053070e-2$	$2.455122601843477e-2$
$(3,1)^\top$	$7.915442225201668e-3$	$5.549250395460341e-3$
$(0,2)^\top$	$-6.889684084070262e-2$	$1.100850009702430e-3$
$(1,2)^\top$	$-7.797901255882132e-3$	$-8.703852796407199e-3$
$(2,2)^\top$	$2.391985596722631e-2$	$-1.107898945797622e-2$
$(3,2)^\top$	$-2.730663197242868e-3$	$-1.824711656717802e-3$
$(0,3)^\top$	$2.786835409915918e-2$	$7.150650562731158e-3$
$(1,3)^\top$	$2.583677812980664e-3$	$3.739425094997325e-3$
$(2,3)^\top$	$-1.022629130739209e-2$	$1.325604608219210e-4$
$(3,3)^\top$	$1.124207929206836e-3$	$-3.153935280982471e-5$
$\beta \in \Lambda_1$	$\widetilde{a}_\beta^{0,1}$	$\widetilde{a}_{\beta+\rho_1}^{1,1}$
0	$3.204127523553855e-1$	$1.561048399423833e-1$
$(1,0)^\top$	$-1.368190023553855e-1$	$2.748891005761668e-2$
$(0,1)^\top$	$1.821862274897458e-1$	$1.420549763613697e-1$
$(1,1)^\top$	$-2.398310248974584e-2$	$1.614814863863025e-2$

primal: Φ_3^{sc}

$\beta \in \Lambda_0$	$\Phi_{5,3}^{sc}$ $\widetilde{a}_\beta^{0,0}$	$\widetilde{\Phi}_{5,3}^{v}$ $\widetilde{a}_\beta^{0,0}$	$\widetilde{a}_{\beta+\rho_1}^{1,0}$
0	$1.248860584413378e+0$	$1.167615178187773e+0$	$1.446572429884526e+0$
$(1,0)^\top$	$-1.072570932069709e-1$	$-7.728925091671431e-2$	$-1.433277093828857e-1$
$(2,0)^\top$	$1.558735171187718e-1$	$1.888546938821553e-1$	$6.085221561513490e-2$
$(3,0)^\top$	$-8.955070805230042e-2$	$-1.187284742137577e-1$	$-5.378123789590539e-2$
$(4,0)^\top$	$2.096572192453899e-2$	$2.860724827395809e-2$	$1.713110069260200e-2$
$(5,0)^\top$	$-1.922667490728697e-3$	$-2.712743619527932e-3$	$-1.621521471208812e-3$
$(0,1)^\top$	$-1.870410495370978e-2$	$-1.403335900393712e-1$	$5.863037052559802e-2$
$(1,1)^\top$	$-6.885758248669444e-2$	$-4.793393782781558e-2$	$-9.535676594603251e-2$
$(2,1)^\top$	$-1.858256869770499e-2$	$4.086806633413162e-2$	$-5.062314323670625e-2$
$(3,1)^\top$	$2.954225407576334e-2$	$8.709039637571915e-3$	$5.538733948680068e-2$
$(4,1)^\top$	$-1.051752726294012e-2$	$-9.153419751946031e-3$	$-1.714419046359280e-2$
$(5,1)^\top$	$8.631799734310942e-4$	$7.727497527436601e-4$	$1.517278021731859e-3$
$(0,2)^\top$	$3.715665709594559e-2$	$2.706775067857388e-3$	$5.569263783068024e-2$
$(1,2)^\top$	$1.316573719691788e-2$	$2.280972899981356e-2$	$1.168667506074336e-2$
$(2,2)^\top$	$-5.143439255359894e-3$	$9.587462117313924e-3$	$-1.596134119595940e-2$
$(3,2)^\top$	$2.172726483773679e-3$	$-7.287080015136010e-3$	$3.790644889919228e-3$
$(4,2)^\top$	$1.945970082387100e-3$	$4.440009723757393e-3$	$3.495881655619283e-3$
$(5,2)^\top$	$4.239569430844405e-5$	$-1.417896096775507e-4$	$-9.646057566259234e-5$
$(0,3)^\top$	$5.383358933710038e-3$	$1.111979655321926e-2$	$8.751383345390871e-4$
$(1,3)^\top$	$-4.372920227797116e-3$	$-6.224201371556257e-3$	$-1.460320384266236e-3$
$(2,3)^\top$	$-4.101113058409931e-3$	$-5.680499646148411e-3$	$-2.424581192953283e-3$
$(3,3)^\top$	$1.771847259905660e-3$	$3.506951089912501e-3$	$-1.007506201115389e-3$
$(4,3)^\top$	$-1.154042970945087e-3$	$-2.442875192961218e-3$	$-5.764645368162608e-4$
$(5,3)^\top$	$3.759640539145489e-5$	$1.537737191437562e-4$	$-9.564997711837430e-5$
$(0,4)^\top$	$-1.128882896364643e-3$	$1.488702290401376e-3$	$-3.338595407112227e-3$
$(1,4)^\top$	$3.278983151430856e-4$	$-9.110977301904477e-4$	$1.156779805136713e-3$
$(2,4)^\top$	$1.624873804498521e-4$	$-8.382304989510840e-4$	$7.937529028566722e-4$
$(3,4)^\top$	$-2.707553531526255e-4$	$9.086381256274376e-4$	$-1.036200932986718e-3$
$(4,4)^\top$	$4.019540677324694e-4$	$9.387935375039568e-5$	$8.755448006994410e-4$
$(5,4)^\top$	$-5.714296199046007e-5$	$2.459604563010142e-6$	$-1.205788721499952e-4$
$(0,5)^\top$	$-1.341465581454279e-4$	$5.767099484458133e-4$	$-5.054014841575678e-4$
$(1,5)^\top$	$6.890501685353420e-5$	$-3.179697641278453e-4$	$2.665717299703359e-4$
$(2,5)^\top$	$8.187897788907335e-5$	$-9.208229619793408e-5$	$2.251497764690581e-4$
$(3,5)^\top$	$-7.554510029609942e-5$	$2.853576033795180e-4$	$-2.719809494541078e-4$
$(4,5)^\top$	$-1.480569881635941e-5$	$-1.962726780249726e-4$	$2.755096560972586e-5$
$(5,5)^\top$	$6.640083442565223e-6$	$3.261216074832735e-5$	$5.409219483771981e-6$
$\beta \in \Lambda_1$	$\widetilde{a}_\beta^{0,1}$	$\widetilde{a}_\beta^{0,1}$	$\widetilde{a}_{\beta+\rho_1}^{1,1}$
0	$6.082108306133408e-1$	$6.547914852818604e-1$	$4.742933247739849e-1$
$(1,0)^\top$	$-1.829061097649758e-1$	$-2.657621840016456e-1$	$-3.326462357924324e-2$
$(2,0)^\top$	$7.469527915163494e-2$	$1.109706987197851e-1$	$5.897129880525828e-2$
$(0,1)^\top$	$-6.561561368690245e-4$	$8.464125900104696e-2$	$-4.621352655426576e-2$
$(1,1)^\top$	$3.554414343527990e-2$	$-5.824546422272656e-2$	$1.125369272173019e-1$
$(2,1)^\top$	$-3.488798729841087e-2$	$-2.639579477832039e-2$	$-6.632340066303617e-2$
$(0,2)^\top$	$-1.048050916980139e-2$	$1.178653391333556e-2$	$-2.881917138803943e-2$
$(1,2)^\top$	$7.080786447207994e-3$	$-2.848396021647917e-2$	$2.604965101234818e-2$
$(2,2)^\top$	$3.399722722593394e-3$	$1.669742630314360e-2$	$2.769520375691254e-3$

primal: Φ_3^v

$\beta \in \Lambda_0$	$\widetilde{\Phi}_{5,3}^v$	
	$\widetilde{a}_\beta^{0,0}$	$\widetilde{a}_{\beta+\rho_1}^{1,0}$
0	$1.376634423200461e + 0$	$1.605808928234782e + 0$
$(1,0)^\top$	$-9.284346344637796e - 2$	$-1.330070581905590e - 1$
$(2,0)^\top$	$1.555481620561534e - 1$	$4.730951249653528e - 2$
$(3,0)^\top$	$-5.387061220097746e - 2$	$-1.412727590628023e - 2$
$(4,0)^\top$	$8.889420404800214e - 3$	$2.540817447257451e - 3$
$(5,0)^\top$	$-5.311302914606342e - 4$	$-1.108718419768552e - 4$
$(0,1)^\top$	$-5.568776915048601e - 2$	$-1.658862256999892e - 1$
$(1,1)^\top$	$-5.071091314326795e - 2$	$-5.072632604906623e - 2$
$(2,1)^\top$	$-2.013751301192182e - 2$	$3.538632534061169e - 2$
$(3,1)^\top$	$1.620116722476630e - 3$	$1.380225574763569e - 3$
$(4,1)^\top$	$-1.214730214968739e - 3$	$-1.639340292750591e - 3$
$(5,1)^\top$	$-1.053313813422444e - 4$	$1.499726721691176e - 4$
$(0,2)^\top$	$-2.844816616315944e - 2$	$1.326803720238974e - 2$
$(1,2)^\top$	$2.299568182253520e - 3$	$6.096150186695251e - 4$
$(2,2)^\top$	$1.180214090634767e - 2$	$-9.013197145961639e - 3$
$(3,2)^\top$	$-3.843814826333664e - 3$	$-2.101228399370152e - 3$
$(4,2)^\top$	$8.993266232835055e - 4$	$8.565629928182265e - 4$
$(5,2)^\top$	$2.163109213159980e - 5$	$-3.100217124791819e - 5$
$(0,3)^\top$	$1.156457264651012e - 2$	$-5.492985386111831e - 3$
$(1,3)^\top$	$-2.263091987592424e - 3$	$-2.860964057243435e - 3$
$(2,3)^\top$	$-5.998153942715757e - 3$	$2.371181297884909e - 3$
$(3,3)^\top$	$1.596067544430546e - 3$	$2.216498626453074e - 3$
$(4,3)^\top$	$-4.625021559537686e - 4$	$-3.030583802434601e - 4$
$(5,3)^\top$	$-1.134533225258886e - 5$	$-3.390434462410624e - 5$
$(0,4)^\top$	$1.288996871424384e - 3$	$2.989783826076559e - 3$
$(1,4)^\top$	$-7.376436340053374e - 4$	$7.119588180794440e - 5$
$(2,4)^\top$	$-3.826773003266109e - 4$	$-1.467211640224730e - 3$
$(3,4)^\top$	$7.535031816069951e - 4$	$1.226063361790190e - 5$
$(4,4)^\top$	$-1.788775468903265e - 4$	$5.526331568170471e - 5$
$(5,4)^\top$	$6.708404089359737e - 5$	$-5.129269305917772e - 7$
$(0,5)^\top$	$-2.778259889168003e - 5$	$7.156217256814941e - 4$
$(1,5)^\top$	$-3.722417815508441e - 5$	$-4.903620157380687e - 4$
$(2,5)^\top$	$-2.879681912346664e - 5$	$-2.844053219285236e - 4$
$(3,5)^\top$	$-2.566176193167613e - 5$	$4.128102343523521e - 4$
$(4,5)^\top$	$-2.053937102133819e - 5$	$-1.366330305028683e - 4$
$(5,5)^\top$	$-3.415495038843051e - 7$	$1.432429179507176e - 5$
$\beta \in \Lambda_1$	$\widetilde{a}_\beta^{0,1}$	$\widetilde{a}_{\beta+\rho_1}^{1,1}$
0	$4.521987792304032e - 1$	$4.880781875790933e - 1$
$(1,0)^\top$	$-1.329449512187649e - 1$	$-1.647601533882240e - 1$
$(2,0)^\top$	$1.930818370711174e - 2$	$1.524397752788076e - 2$
$(0,1)^\top$	$7.891896646919766e - 2$	$1.484235127815599e - 2$
$(1,1)^\top$	$-2.501534030258234e - 2$	$6.646628606360538e - 2$
$(2,1)^\top$	$1.262469414588470e - 2$	$-1.478031702926138e - 2$
$(0,2)^\top$	$6.815624730985620e - 3$	$3.452406009121726e - 2$
$(1,2)^\top$	$4.577075561319153e - 3$	$-2.378469798030484e - 2$
$(2,2)^\top$	$2.797973535820227e - 3$	$3.451311717212584e - 3$

Bibliography

[1] ISO/IEC JTC1 10918-x, *Information technology – Digital compression and coding of continuous-tone still images*, 1994.

[2] ISO/IEC JTC 1 15444-x, *Information technology – JPEG 2000 image coding system*, 2000.

[3] ANSI/IEEE Std 754-1985, *IEEE standard for binary floating-point arithmetic*, 1985.

[4] F. S. Acton, *Numerical methods that work*, pp. 464–467, Mathematical Association of America, Washington D.C., 1990.

[5] B. K. Alpert, *A class of bases in L_2 for the sparse representation of integral operators*, SIAM J. Math. Anal. **24** (1993), no. 1, 246–262.

[6] M. Antonini, M. Barlaud, I. Daubechies, and P. Mathieu, *Image coding using wavelet transform*, IEEE Trans. Image Process. **1** (1992), no. 2, 205–220.

[7] A. E. Bell and L. R. Iyer, *Improving image compression performance with balanced multiwavelets*, Conference Record of the Thirty-Fifth Asilomar Conference on Signals, Systems and Computers (Pacific Grove, CA, USA), vol. 1, 2001, pp. 773–777.

[8] A. E. Bell and M. B. Martin, *New image compression techniques using multiwavelets and multiwavelet packets*, IEEE Trans. Image Process. **10** (2001), no. 4, 500–510.

[9] A. E. Bell and S. Rout, *Color image compression: multiwavelets vs. scalar wavelets*, Proceedings of the IEEE International Conference on Acoustics, Speech, and Signal Processing (ICASSP '02), vol. 4, 2002, pp. IV–3501–IV–3504.

[10] C. Cabrelli, C. Heil, and U. Molter, *Accuracy of lattice translates of several multidimensional refinable functions*, J. Approx. Theory **95** (1998), 5–52.

[11] _____, *Accuracy of several multidimensional refinable distributions*, J. Fourier Anal. Appl. **6** (2000), no. 5, 483–502.

[12] _____, *Self–similarity and multiwavelets in higher dimensions*, Mem. Am. Math. Soc. **170** (2004), no. 807, 82 p.

[13] P. J. Cameron, *Permutation groups*, London Mathematical Society Student Texts, vol. 45, Cambridge University Press, 1999.

[14] A. S. Cavaretta, W. Dahmen, and C. A. Micchelli, *Stationary subdivision*, Mem. Am. Math. Soc. **93** (1991), no. 453, 186 p.

[15] M. Charina, C. Conti, and T. Sauer, *Regularity of multivariate vector subdivision schemes*, Numer. Algorithms **39** (2005), no. 1–3, 97–113.

[16] D. R. Chen, R. Q. Jia, and S. D. Riemenschneider, *Convergence of vector subdivision schemes in Sobolev spaces*, Appl. Comput. Harmon. Anal. **12** (2002), no. 1, 128–149.

[17] O. Christensen, *An Introduction to Frames and Riesz Bases*, Birkhäuser, 2003.

[18] C. K. Chui and Q. T. Jiang, *Multivariate balanced vector–valued refinable functions*, Modern developments in multivariate approximation. Proceedings of the 5th international conference, Witten–Bommerholz, Germany, September 22–27 (W. Haussmann, K. Jetter, M. Reimer, and J. Stöckler, eds.), Int. Ser. Numer. Math., vol. 145, Birkhäuser, 2003, pp. 71–102.

[19] _____, *Balanced multi-wavelets in* $\mathbb{R}^s$, Math. Comput. **74** (2005), 1323–1344.

[20] _____, *Matrix-valued symmetric templates for interpolatory surface subdivisions: I. regular vertices*, Appl. Comput. Harmon. Anal. **19** (2005), no. 3, 303–339.

[21] _____, *Matrix-valued subdivision schemes for generating surfaces with extraordinary vertices*, Comput. Aided Geom. Design **23** (2006), 419–438.

[22] C. K. Chui and C. Li, *A general framework of multivariate wavelets with duals*, Appl. Comput. Harmon. Anal. **1** (1994), no. 4, 368–390.

[23] A. Cohen, *Numerical analysis of wavelet methods*, Studies in Mathematics and its Applications, vol. 32, North–Holland, Amsterdam, 2003.

[24] A. Cohen and I. Daubechies, *Non–separable bidimensional wavelet bases*, Rev. Mat. Iberoamericana **9** (1993), 51–137.

[25] A. Cohen, I. Daubechies, and J.-C. Feauveau, *Biorthogonal bases of compactly supported wavelets*, Commun. Pure Appl. Math. **45** (1992), 485–560.

[26] A. Cohen, I. Daubechies, and G. Plonka, *Regularity of refinable function vectors*, J. Fourier Anal. Appl. **3** (1997), 295–324.

[27] R.R. Coifman and D.L. Donoho, *Translation invariant denoising*, Lecture Notes in Statistics, vol. 103, Springer Verlag, New York, 1995.

[28] C. Conti and G. Zimmermann, *Interpolatory rank-1 vector subdivision schemes*, Comput. Aided Geom. Des. **21** (2004), no. 4, 341–351.

[29] S. Dahlke, W. Dahmen, and V. Latour, *Smooth refinable functions and wavelets obtained by convolution products*, Appl. Comput. Harmon. Anal. **2** (1995), no. 1, 68–84.

[30] S. Dahlke, K. Gröchenig, and P. Maass, *A new approach to interpolating scaling functions*, Applicable Analysis **72** (1999), no. 3-4, 485–500.

[31] S. Dahlke and P. Maass, *Interpolating refinable functions and wavelets for general scaling matrices*, Numer. Funct. Anal. Optim. **18** (1997), no. 5&6, 521–539.

[32] S. Dahlke, P. Maass, and G. Teschke, *Interpolating scaling functions with duals*, J. Comput. Anal. Appl. **6** (2004), no. 1, 19–29.

[33] W. Dahmen and C. A. Micchelli, *Using the refinement equation for evaluating integrals of wavelets*, SIAM J. Numer. Anal. **30** (1993), no. 2, 507–537.

[34] ———, *Biorthogonal wavelet expansions*, Constr. Approx. **13** (1997), 294–328.

[35] USC-SIPI Image Database, `http://sipi.usc.edu/database/`.

[36] I. Daubechies, *Orthonormal bases of compactly supported wavelets*, Commun. Pure Appl. Math. **41** (1988), no. 7, 909–996.

[37] ———, *Ten lectures on wavelets*, CBMS–NSF Regional Conference Series in Applied Math., vol. 61, SIAM, Philadelphia, 1992.

[38] I. Daubechies, A. Grossmann, and Y. Meyer, *Painless nonorthogonal expansions*, J. Math. Phys. **27** (1986), 1271–1283.

[39] C. de Boor, R. DeVore, and A. Ron, *On the construction of multivariate (pre-) wavelets*, Constr. Approx. **9** (1993), no. 2, 123–166.

[40] _____, *The structure of finitely generated shift-invariant spaces in $L_2(\mathbb{R}^d)$*, J. Funct. Anal. **119** (1994), no. 1, 37–78.

[41] C. de Boor, K. Höllig, and S. D. Riemenschneider, *Box splines*, Applied Mathematical Sciences, vol. 98, Springer, New York, 1993.

[42] J. Derado, *Nonseparable, compactly supported interpolating refinable functions with arbitrary smoothness*, Appl. Comput. Harmon. Anal. **10** (2001), no. 2, 113–138.

[43] G. Deslauriers, J. Dubois, and S. Dubuc, *Multidimensional iterative interpolation*, Can. J. Math. **43** (1991), no. 43, 297–312.

[44] P. Deuflhard and A. Hohmann, *Numerische mathematik I*, Walter de Gruyter, Berlin–New York, 2002.

[45] R. DeVore, B. Jawerth, and B. Lucier, *Image compression through wavelet transform coding*, IEEE Trans. Information Theory **38** (1992), no. 2, 719–746, Special issue on Wavelet Transforms and Multiresolution.

[46] R. DeVore, B. Jawerth, and V. Popov, *Compression of wavelet decompositions*, Am. J. Math. **114** (1992), 737–785.

[47] T. Flaherty and Y. Wang, *Haar–type multiwavelet bases and self–affine multi–tiles*, Asian J. Math. **3** (1999), no. 2, 387–400.

[48] D. Gabor, *Theory of communication*, J. Inst. Elect. Engng. **93** (1946), 429–457.

[49] J. Geronimo, D. Hardin, and P. R. Massopust, *Fractal functions and wavelet expansions based on several functions*, J. Approximation Theory **78** (1994), 373–401.

[50] T. N. T. Goodman and S. L. Lee, *Wavelets of multiplicity r*, Trans. Amer. Math. Soc. **342** (1994), no. 1, 307–324.

[51] T. N. T. Goodman, S. L. Lee, and W. S. Tang, *Wavelets in wandering subspaces*, Trans. Amer. Math. Soc. **338** (1993), no. 2, 639–654.

[52] K. Gröchenig, A. Haas, and A. Raugi, *Self-affine tilings with several tiles, I*, Appl. Comput. Harmon. Anal. **7** (1999), no. 1, 211–238.

[53] K. Gröchenig and W. R. Madych, *Multiresolution analysis, Haar bases, and self-similar tilings of* $\mathbb{R}^n$, IEEE Trans. Inf. Theory **38** (1992), no. 2/II, 556–568.

[54] A. Grossmann and J. Morlet, *Decomposition of Hardy functions into square integrable wavelets of constant shape*, SIAM J. Math. Anal. **15** (1984), 723–736.

[55] A. Haar, *Zur Theorie der orthogonalen Funktionensysteme*, Math. Annalen **69** (1910), 331–371.

[56] B. Han, *Approximation properties and construction of Hermite interpolants and biorthogonal multiwavelets*, J. Approximation Theory **110** (2001), no. 1, 18–53.

[57] _____, *Symmetry property and construction of wavelets with a general dilation matrix*, Linear Algebra Appl. **353** (2002), no. 1–3, 207–225.

[58] _____, *Computing the smoothness exponent of a symmetric multivariate refinable function*, SIAM J. Matrix Anal. Appl. **24** (2003), no. 3, 693–714.

[59] _____, *Vector cascade algorithms and refinable function vectors in Sobolev spaces*, J. Approximation Theory **124** (2003), no. 1, 44–88.

[60] _____, *Symmetric multivariate orthogonal refinable functions*, Appl. Comput. Harmon. Anal. **17** (2004), no. 3, 277–292.

[61] B. Han and R.-Q. Jia, *Multivariate refinement equations and convergence of subdivision schemes*, SIAM J. Math. Anal. **29** (1998), 1177–1199.

[62] _____, *Optimal interpolatory subdivision schemes in multidimensional spaces*, SIAM J. Numer. Anal. **36** (1998), 105–124.

[63] _____, *Quincunx fundamental refinable functions and quincunx biorthogonal wavelets*, Math. Comput. **71** (2002), no. 237, 165–196.

[64] B. Han, B. Piper, and P.-Y. Yu, *Multivariate refinable Hermite interpolants*, Math. Comput. **73** (2004), no. 248, 1913–1935.

[65] C. Heil, G. Strang, and V. Strela, *Approximation by translates of refinable functions*, Numer. Math. **73** (1996), 75–94.

[66] C. Heil and D. F. Walnut (eds.), *Fundamental papers in wavelet theory*, Princeton University Press, Princeton, 2006.

[67] L. Hervé, *Multi-resolution analysis of multiplicity d: Applications to dyadic interpolation*, Appl. Comput. Harmon. Anal. **1** (1994), no. 4, 299–315.

[68] H. Ji, S. D. Riemenschneider, and Z. Shen, *Multivariate compactly supported fundamental refinable functions, duals, and biorthogonal wavelets*, Stud. Appl. Math. **102** (1999), no. 2, 173–204.

[69] R. Q. Jia, *Shift–invariant spaces and linear operator equations*, Isr. J. Math. **103** (1998), 259–288.

[70] R. Q. Jia and Q. T. Jiang, *Approximation power of refinable vectors of functions*, Wavelet analysis and applications. Proceedings of an international conference, Guangzhou, China, November 15-20, 1999 (D. Deng, D. Huang, R. Q. Jia, W. Lin, and J. Wang, eds.), Stud. Adv. Math., vol. 25, Amer. Math. Soc., Providence, RI, 2002, pp. 155–178.

[71] ———, *Spectral analysis of the transition operator and its applications to smoothness analysis of wavelets*, SIAM J. Matrix Anal. Appl. **24** (2003), no. 4, 1071–1109.

[72] R. Q. Jia and C. A. Micchelli, *On linear independence for integer translates of a finite number of functions*, Proc. Edinb. Math. Soc. **36** (1992), 69–85.

[73] R. Q. Jia, S. D. Riemenschneider, and D. X. Zhou, *Smoothness of multiple refinable functions and multiple wavelets*, SIAM J. Matrix Anal. Appl. **21** (1999), 1–28.

[74] R. Q. Jia and Z. Shen, *Multiresolution and wavelets*, Proc. Edinb. Math. Soc. **37** (1994), 271–300.

[75] Q. T. Jiang, *Multivariate matrix refinable functions with arbitrary matrix dilation*, Trans. Amer. Math. Soc. **351** (1999), 2407–2438.

[76] Q. T. Jiang, P. Oswald, and S. D. Riemenschneider, *$\sqrt{3}$-subdivision schemes: maximal sum rule orders*, Constr. Approx. **19** (2003), 437–463.

[77] Q. T. Jiang and Z. Shen, *On existence and weak stability of matrix refinable functions*, Constr. Approx. **15** (1999), 337–353.

[78] B. Kessler, *A construction of compactly supported biorthogonal scaling vectors and multiwavelets on $\mathbb{R}^2$*, J. Approximation Theory **117** (2002), 229–254.

[79] K. Koch, *Interpolating scaling vectors*, Int. J. Wavelets Multiresolut. Inf. Process. **3** (2005), no. 3, 1–29.

[80] _____, *Multivariate orthonormal interpolating scaling vectors*, Appl. Comput. Harmon. Anal. (2006), to appear.

[81] _____, *Multivariate symmetric interpolating scaling vectors*, Tech. Report 145, Preprint Series DFG-SPP 1114, 2006, submitted.

[82] W. Lawton, S. L. Lee, and Zuowei Shen, *An algorithm for matrix extension and wavelet construction*, Math. Comput. **65** (1996), no. 214, 723–737.

[83] J. Lebrun and M. Vetterli, *Balanced multiwavelets: Theory and design*, IEEE Trans. on Signal Processing **46** (1998), no. 4, 1119–1125.

[84] _____, *High order balanced multiwavelets*, In Proc. IEEE Int. Conf. Acoust., Speech, Signal Processing (ICASSP) (Seattle), 1998, pp. 12–15.

[85] M. Lindemann, *Nonlinear approximation methods using wavelets and multiwavelets for general scaling matrices*, Dissertation, Zentrum für Technomathematik, Universität Bremen, 2005.

[86] A. K. Louis, P. Maass, and A. Rieder, *Wavelets. Theory and applications*, John Wiley, Chichester, 1997.

[87] S. Mallat, *Multiresolution approximation and wavelet orthonormal bases of $L_2(\mathbb{R}^d)$*, Trans. Amer. Math. Soc. **315** (1989), 69–87.

[88] _____, *A wavelet tour of signal processing*, 2. ed., Academic Press, San Diego, 1999.

[89] Y. Meyer, *Principe d'incertitude, bases hilbertiennes et algèbres d'opérateurs*, Sémin. Bourbaki **38** (1985/86), no. 662, 209–223.

[90] _____, *Ondelettes, fonctions splines et analyses graduées*, Rend. Semin. Mat. Univ. Politec. Torino **45** (1987), 1–42.

[91] _____, *Ondelettes et opérateurs I: Ondelettes*, Hermann, 1990.

[92] C. A. Micchelli, *Interpolatory subdivision schemes and wavelets*, J. Approximation Theory **86** (1996), no. 1, 41–71.

[93] C. A. Micchelli and T. Sauer, *Regularity of multiwavelets*, Adv. Comput. Math. **7** (1997), 455–545.

[94] _____, *Continuous refinable functions and self similarity*, Vietnam Journal of Mathematics **31** (2003), 449–464.

[95] J. A. Nelder and R. Mead, *A simplex method for function minimization*, Computer J. **7** (1965), 308–313.

[96] A. M. N. Niklasson, H. Röder, and C. J. Tymczak, *Separable and nonseparable multiwavelets in multiple dimensions*, J. Comput. Phys. **175** (2002), 363–397.

[97] W. A. Pearlman and A. Said, *A new fast and efficient image codec based on set partitioning in hierarchical trees*, IEEE Trans. on Circ. and Syst. for Video Tech. **6** (1996), no. 3, 243–250.

[98] G. Plonka, *Approximation order provided by refinable function vectors*, Constr. Approx. **13** (1997), 221–244.

[99] G. Plonka and A. Ron, *A new factorization technique of the matrix mask of univariate refinable functions*, Numer. Math. **87** (2001), 555–595.

[100] M. J. D. Powell, *An efficient method for finding the minimum of a function of several variables without calculating derivatives*, Comput. J. **7** (1964), 155–162.

[101] S. D. Riemenschneider and Z. Shen, *Multidimensional interpolatory subdivision schemes*, SIAM J. Numer. Anal. **34** (1997), no. 6, 2357–2381.

[102] ———, *Construction of compactly supported biorthogonal wavelets in $L_2(\mathbb{R}^d)$ II*, Wavelet Applications in Signal and Image Processing VII (A. Aldroubi, A. Laine, and M. Unser, eds.), Proceedings of the SPIEE, vol. 3813, 1999, pp. 264–272.

[103] A. M. C. Ruedin, *A nonseparable multiwavelet for edge detection*, Wavelets: Applications in Signal and Image Processing X (A. Aldroubi, A. Laine, and M. Unser, eds.), Proceedings of the SPIEE, vol. 5207, 2003, pp. 700–709.

[104] N. Saßmannshausen, *Allgemeine mehrdimensionale Wavelet-Theorie und Spektraleigenschaften des Transferoperators*, Ph.D. thesis, Fachbereich Mathematik und Informatik, Philipps-Universität Marburg, 2002.

[105] T. Sauer, *Stationary vector subdivision — quotient ideals, differences and approximation power*, Rev. R. Acad. Cien. Serie A. Mat. **96** (2002), 257–277.

[106] ———, *Polynomial interpolation in several variables: lattices, differences, and ideals*, Tech. report, Justus-Liebig-Universität Gießen, 2006, submitted.

[107] W. Schempp and B. Dreseler, *Einführung in die harmonische Analyse*, B. G. Teubner Verlag, Stuttgart, 1980.

[108] I. W. Selesnick, *Multiwavelet bases with extra approximation properties*, IEEE Trans. on Signal Processing **46** (1998), no. 11, 2898–2909.

[109] ———, *Interpolating multiwavelet bases and the sampling theorem*, IEEE Trans. on Signal Processing **47** (1999), no. 6, 1615–1621.

[110] J. M. Shapiro, *Embedded image coding using zerotrees of wavelet coefficients*, IEEE Trans. on Signal Processing **41** (1993), no. 12, 3445–3462.

[111] Z. Shen, *Refinable function vectors*, SIAM J. Math. Anal. **29** (1998), 235–250.

[112] G. Strang and T. Nguyen, *Wavelets and filter banks*, Wellesley-Cambridge Press, Wellesley, 1996.

[113] V. Strela, P. N. Heller, G. Strang, P. Topiwala, and C. Heil, *The application of multiwavelet filter banks to image processing*, IEEE Trans. Image Process. **8** (1999), 548–563.

[114] V. Strela and A. T. Walden, *Signal and image denoising via wavelet thresholding: Orthogonal and biorthogonal, scalar and multiple wavelet transforms*, Nonlinear and Nonstationary Signal Processing (W. F. Fitzgerald, R. L. Smith, A. T. Walden, and P. C. Young, eds.), Cambridge University Press, 2001.

[115] J.-O. Strömberg, *A modified Franklin system and higher-order spline systems on $\mathbb{R}^d$ as unconditional bases for Hardy spaces*, Harmonic analysis, Conf. in Honor A. Zygmund, Chicago 1981, 1983, pp. 475–494.

[116] T. Strutz, *Bilddatenkompression*, Vieweg, Braunschweig-Wiesbaden, 2000.

[117] J. Walker, *Fractal food*, http://www.fourmilab.ch/images/Romanesco/.

[118] P. Wojtaszczyk, *A mathematical introduction to wavelets*, Cambridge University Press, 1997.

[119] X.-G. Xia, J. S. Geronimo, D. P. Hardin, and B. Q. Suter, *Design of prefilters for discrete multiwavelet transforms*, IEEE Trans. on Signal Processing **44** (1996), no. 1, 25–35.

[120] X.-G. Xia and Z. Zhang, *On sampling theorem, wavelets and wavelet transforms*, IEEE Trans. on Signal Processing **41** (1993), no. 12, 3524–3535.